STRUCTURES IN THE UNIVERSE BY EXACT METHODS

Formation, Evolution, Interactions

As the structures in our Universe are mapped out on ever larger scales, and with increasing detail, the use of inhomogeneous models is becoming an essential tool for analysing and understanding them.

This book reviews a number of important developments in the application of inhomogeneous solutions of Einstein's field equations to cosmology. It shows how inhomogeneous models can be used to study the evolution of structures such as galaxy clusters and galaxies with central black holes, and to account for cosmological observations such as supernova dimming, the cosmic microwave background, baryon acoustic oscillations or the dependence of the Hubble parameter on redshift within classical general relativity.

Whatever 'dark matter' and 'dark energy' turn out to be, inhomogeneities exist on many scales and need to be investigated by all appropriate methods. This book will be of great value to astrophysicists and cosmologists, from graduate students to academic researchers.

KRZYSZTOF BOLEJKO is an Assistant Professor at the Nicolaus Copernicus Astronomical Center, Polish Academy of Sciences, Poland. His area of research is cosmology, in particular applications of inhomogeneous cosmological models, study of light propagation, and backreaction effects.

ANDRZEJ KRASIŃSKI is an Associate Professor at Nicolaus Copernicus Astronomical Center, Polish Academy of Sciences, Poland. He is currently doing research in gravitation theory, in particular exact solutions of Einstein's equations and inhomogeneous cosmological models. He is also a member of the editorial board of the *GRG Journal*, where he is the editor of the *"Golden Oldies"* series of reprints of classical papers on relativity.

CHARLES HELLABY is an Associate Professor in the Department of Mathematics and Applied Mathematics, University of Cape Town, South Africa. His research interests are in cosmology and relativity, particularly properties and applications of inhomogeneous metrics, and relating observations to cosmological models.

MARIE-NOËLLE CÉLÉRIER is a Researcher at the Laboratoire Univers et Théories (LUTH), Observatoire de Paris-Meudon, France. Her two main fields of interest are scale relativity and inhomogeneous cosmology.

CAMBRIDGE MONOGRAPHS ON MATHEMATICAL PHYSICS

General editors: P. V. Landshoff, D. R. Nelson, S. Weinberg

R. Penrose and W. Rindler *Spinors and Space-Time Volume 1: Two-Spinor Calculus and Relativistic Fields*[†]
R. Penrose and W. Rindler *Spinors and Space-Time Volume 2: Spinor and Twistor Methods in Space-Time Geometry*[†]
S. Pokorski *Gauge Field Theories,* 2nd edition[†]
J. Polchinski *String Theory Volume 1: An Introduction to the Bosonic String*
J. Polchinski *String Theory Volume 2: Superstring Theory and Beyond*
V. N. Popov *Functional Integrals and Collective Excitations*[†]
R. J. Rivers *Path Integral Methods in Quantum Field Theory*[†]
R. G. Roberts *The Structure of the Proton: Deep Inelastic Scattering*[†]
C. Rovelli *Quantum Gravity*[†]
W. C. Saslaw *Gravitational Physics of Stellar and Galactic Systems*[†]
M. Shifman and A. Yung *Supersymmetric Solitons*
H. Stephani, D. Kramer, M. MacCallum, C. Hoenselaers and E. Herlt *Exact Solutions of Einstein's Field Equations,* 2nd edition
J. Stewart *Advanced General Relativity*[†]
T. Thiemann *Modern Canonical Quantum General Relativity*
D. J. Toms *The Schwinger Action Principle and Effective Action*
A. Vilenkin and E. P. S. Shellard *Cosmic Strings and Other Topological Defects*[†]
R. S. Ward and R. O. Wells, Jr *Twistor Geometry and Field Theory*[†]
J. R. Wilson and G. J. Mathews *Relativistic Numerical Hydrodynamics*

[†] Issued as a paperback.

Structures in the Universe by Exact Methods

Formation, Evolution, Interactions

KRZYSZTOF BOLEJKO

N. Copernicus Astronomical Center, Polish Academy of Sciences, Poland

ANDRZEJ KRASIŃSKI

N. Copernicus Astronomical Center, Polish Academy of Sciences, Poland

CHARLES HELLABY

University of Cape Town, South Africa

MARIE-NOËLLE CÉLÉRIER

Laboratoire Univers et Théories (LUTH), Observatoire de Paris, France

CAMBRIDGE
UNIVERSITY PRESS

CAMBRIDGE UNIVERSITY PRESS

Cambridge, New York, Melbourne, Madrid, Cape Town, Singapore, São Paulo, Delhi

Cambridge University Press
The Edinburgh Building, Cambridge CB2 8RU, UK

Published in the United States of America by Cambridge University Press, New York

www.cambridge.org
Information on this title: www.cambridge.org/9780521769143

First published 2010

Printed in the United Kingdom at the University Press, Cambridge

A catalogue record for this publication is available from the British Library

Library of Congress Cataloguing in Publication data
Structures in the universe by exact methods : formation, evolution,
interactions / Krzysztof Bolejko ... [et al.].
p. cm. – (Cambridge monographs on mathematical)
Includes bibliographical references and index.
ISBN 978-0-521-76914-3 (hardback)
1. Cosmology–Mathematics. 2. Einstein field equations. 3. Inhomogeneous materials.
I. Bolejko, Krzysztof, 1980– II. Title. III. Series.
QB981.S87 2009
523.1 – dc22   2009020680

ISBN 978-0-521-76914-3 Hardback

Contents

II APPLICATIONS OF THE MODELS IN COSMOLOGY

Foreword

Cosmology, from its very birth, has been a science beset by insufficient data and ill-founded tenets. Faced with extreme difficulties in collecting observational data, it used to rely on simplifying working assumptions. Those assumptions had the tendency to evolve into dogmas or into elaborate theoretical constructs, immersed in which their practitioners all too easily forgot about these shaky foundations. It is enough to recall the original cosmological paper of Einstein from 1917, who, swayed by prevailing beliefs, was absolutely sure that the Universe must be spatially uniform and unchanging in time – so much so that he preferred to modify his freshly created theory rather than say that it contradicts the astronomical dogma. Another instructive example is the steady state theory. It had been very much in vogue for about 20 years, before it was proved wrong by a single discovery, that of the cosmic microwave background (CMB) radiation. It was glamorous and successful, in the eyes of its proponents, even though it relied on an assumption that is drastically at variance with laboratory physics (continuous creation of matter particles out of nothing).

In spite of such cautionary examples, this tendency continues. Today we are told that the Universe is perfectly homogeneous 'at sufficiently large scales', although the actual size of those 'scales' is not better defined than as 'a few' hundred Mpc. We are told that, consequently, the Friedmann–Lemaître–Robertson–Walker (FLRW) models[1] are the perfect models of the Universe that do not need any modification other than the perturbations needed to account for the formation of structures. By sticking to the FLRW models we are led to believe that more than 95% of the energy density of our Universe consists of entities that no-one has yet seen in any laboratory – aptly called 'dark matter' and 'dark energy'. We are told that the Universe had to go through a phase of exponential expansion called inflation, to 'explain' its extreme homogeneity today. Almost everybody happily believes this, forgetting that the theory postulating this

---

[1] The term 'FLRW models' will be used to denote the general Robertson–Walker class of metrics, with an unspecified scale factor, and solutions of Einstein's equations within this class with a given equation of state $p(\rho)$. The term 'Friedmann models' will denote the dust ($p = 0$) Robertson–Walker models, also when $\Lambda \neq 0$.

phenomenon (I) uses scalar fields which up to now have never been either observed or theoretically characterised; (II) cannot explain how this exponential expansion stopped; (III) has relied on explicit models that are homogeneous from the very beginning, so the process of homogenisation has in fact never been demonstrated; (IV) requires that the duration of the inflationary phase is fine tuned so that the initial variations in the energy density of the inflating region are smoothed out at the end.

Our book aims at showing that good old experimentally verified gravitation theory has not yet been taken to its limits in application to cosmology. Well-established physics can explain several of the phenomena observed by cosmologists without introducing highly speculative elements, like dark matter, dark energy, exponential expansion at densities never attained in any experiment, and the like. We are aware that at some point the Einstein theory may need modification or generalisation. However, we believe that we should undertake such changes only when we really hit a brick wall – i.e. encounter difficulties that cannot be overcome in any other way. As long as the existing theory can explain things, modifications should be laid aside. This has been a good practice in physics before the modern methods of marketing entered our science. This is not to say speculative models should not be explored, rather that they should not displace well-established physics until their theoretical and observational foundations are properly established.

We expect our readers to be well versed in the relativity theory. Also, our book is not intended to be a textbook on relativistic cosmology – we assume quite a bit of prior expertise also in this field. Readers who need an introduction to these subjects will have to consult other sources. There are many good textbooks on relativity on the market, and also several textbooks on cosmology, although most of the latter are strongly biased toward the modern ideas which we criticise. Readers in need of a quick introduction may wish to use the textbook by Plebański and Krasiński (2006) – it may be an easier reading because it has a similar style, the same conventions and the same notation as the present book, and parts of the introductory material here are discussed there at more length. In particular, this applies to the Robertson–Walker models that are not given any introduction here.

This book is mostly based on our own investigations, but we did pay attention to related work by other authors. However, we have included only the applications of exact solutions. Although the averaging methods seem to be very promising in dealing with inhomogeneities, the averaging procedure still lacks a generally accepted methodology. Many interesting aspects of averaging are not uniquely defined and need to be compared with exact solutions. So, while we are looking at this subject with sympathy, we must refer our readers to works by other authors, such as those reported in the articles by

Buchert (2000, 2001); Buchert and Carfora (2002, 2003); Wiltshire (2007a,b), and in the reviews by Räsänen (2006b) and Buchert (2008), and references cited therein.

Krzysztof Bolejko
Andrzej Krasiński
Charles Hellaby
Marie-Noëlle Célérier
October 2008

Acknowledgements

This book grew out of a project carried out by K.B. and A.K., partly supported by the Polish Ministry of Science and Higher Education under grants 1 P03B 075 29 and N203 018 31/2873. K.B. would like to thank Peter and Patricia Gruber and the International Astronomical Union for the PPGF Fellowship. In 2008 part of this research was carried out at the University of Melbourne, thus K.B. would like to thank people from the School of Physics, especially Stuart Wyithe, Rachel Webster and Russell Walsh for their hospitality.

1
The purpose of this book

For many years after its appearance, general relativity (GR) was regarded as an exotic extension of Newtonian gravity, that was only necessary for high-precision measurements in the Solar System and for describing the expansion of the Universe. However, the increasing precision of physical and astronomical measurement is transforming GR into an indispensable tool, and not merely a small correction to Newton's theory.

It is commonly stated that we have entered the era of precision cosmology, in which a number of important observations have reached a degree of precision, and a level of agreement with theory, that is comparable with many Earth-based physics experiments. One of the consequences of this advance is the need to examine at what point our usual, well-worn assumptions begin to compromise the accuracy of our models, and whether more general theoretical methods are needed to maintain calculational accuracy. Historically, each advance in astronomical measurement has produced many new discoveries, and revealed more of the structure of the cosmos, such as voids, walls, filaments, etc. As we map out the Universe around us – its mass distribution and flow patterns – in ever greater detail, the nonlinear behaviour of cosmic structures will become increasingly apparent, and the methods of inhomogeneous cosmology will come into their own. Inhomogeneous solutions of Einstein's field equations provide models of both small and large structures that are fully nonlinear.

It is widely assumed that the Universe, when viewed on a large enough scale, is homogeneous and can be described by an FLRW model. The successes of the Concordance model are built on using a spatially homogeneous and isotropic background metric combined with first-order perturbation theory. Although this assumption has been appropriate up to now, and underlies many important developments in cosmology, it is not the whole story. The modern successes have led to considerable hubris and overconfidence that we have pinned down the matter content and evolution of the key epochs of the post-grand unification Universe, even though all admit we don't know the underlying physics of dark matter and dark energy. Therefore, despite these successes, we must not lose sight of the fact that the present-day Universe is actually very inhomogeneous. If we keep this fact out of our minds, then we ignore knowledge and techniques that will be essential in understanding the real Universe and its multitudinous components in the era of precision cosmology. The relationship between a lumpy universe and an averaged homogeneous one is still not all that well understood, though there are some promising investigations (Buchert, 2000, 2001; Buchert and Carfora, 2002, 2003; Räsänen, 2006; Buchert, 2008; Ellis, 2008; Leith, Ng, and Wiltshire, 2008). The assumption of homogeneity – so essential in developing the basics of cosmology – must now be considered just a zeroth-order approximation; and similarly linear perturbation theory a first-order approximation, whose domain of validity is an early, nearly homogeneous Universe.

The use of perturbations relies on two myths that we wish to briefly discuss here:

1. Since for galaxy clusters the gravitational potential ϕ obeys $\phi/c^2 \ll 1$ (typically $\phi/c^2 \approx 10^{-5}$), the spacetime is nearly flat and linear perturbations work to sufficient accuracy.

2. Where linear perturbations become unsatisfactory, one can go to higher orders of perturbation.

This is how these myths disagree with reality:

1. The fact is that the quantity ϕ/c^2 is a measure of the so-called curvature contrast (perturbation of the 3-dimensional curvature of space divided by the local value of the curvature). But in order to remain safely within the 'linear regime', *both* the curvature contrast and the density contrast, $\Delta\rho/\rho$, must remain small. Moreover, the curvature contrast is a very imprecise indicator of the quality of approximation, since, for example in the Lemaître–Tolman model with a negative curvature background, it is decreasing with time, irrespective of the initial conditions (see Sec. 18.10 in Plebański and Krasiński, 2006), while the density contrast is increasing. Thus, results of such approximate calculations cannot be safely extrapolated into the future. At present, for various galaxy clusters $\Delta\rho/\rho$ is contained between about 5 and about 4000 (see Table 4.1 here), and still much more for smaller structures, so we are already outside the range where the series of approximations can be expected to converge to an exact result. If an exact calculation gives a result that is close to one obtained earlier by a perturbative method, then this is a confirmation that should be welcomed rather than criticised by saying 'it was obvious that the result had to be close to the earlier-calculated one, so the exact treatment was needless.'

2. Higher-order perturbations can improve the accuracy achieved in the first order only when we are still within the range of convergence of the series of approximations (or, in astronomical terminology, within the 'linear regime'). When we are outside that range, second-order corrections will turn out to be larger than the first-order result, and thus worthless. Indeed, second-order calculations are so complex, and involve so many terms, that investigations must focus on particular phenomena and set all other unrelated terms to zero.

3. Although various assertions are made about the requirements for perturbation theory to be valid, we are not aware of any proof of the domain of validity at linear or any other order, nor any proof of convergence. There do seem to be significant difficulties and uncertainties with the method, as discussed in the

following examples. Van Elst *et al.* (1997) show that most irrotational silent cosmological models have a linearisation instability. Bruna and Girbau (1999, 2005) analysed the linearisation stability of Robertson–Walker models, and showed the closed model is not linearisation stable. Notari (2006) suggests that, due to the self-gravitation of perturbations, their contribution at the present day has grown much larger than expected from standard perturbation theory. In a brief review of nonlinear methods for cosmological perturbations, Matarrese (1996) concludes that there are effects that violate the standard wisdom about when relativistic effects are important. Crocce and Scoccimarro (2006) investigate terms up to third order in a perturbation expansion, and they show that terms of successive orders can have different signs, and that second- and third-order terms can exceed first-order terms at larger wavelengths (see their Fig. 1). To address this problem, they introduce renormalised perturbation theory, and though in the examples given the new approach keeps all its terms positive, and higher-order terms contribute at ever smaller scales, the convergence of the renormalised series is also unproven. Losic and Unruh (2008) and Unruh and Losic (2008) find that second-order perturbations could become stronger than first order in models of slow-roll inflation. In the 'nonlinear regime' there is just no escape from exact methods.

On the other hand, the observational data do allow plenty of scope for explanations based on inhomogeneous models. This fact is often hidden under layers of vigorous advertising for traditional methods, but is slowly coming to light – see the broad and impartial review by Sarkar (2008). Similar comments, although strongly biased toward averaging techniques seen as the ultimate best method, can be found in the papers by Wiltshire (2007).

In fact, recent works have tried to combine the averaging method 'à la Buchert' with cosmological linear perturbation theory, claimed to be valid at scales where nonlinear evolution is supposed to enter into play, to study the 'backreaction' of inhomogeneities on a 'homogeneous background' (Li and Schwarz, 2007, 2008). This shows that the cosmological community is becoming aware that the effects of inhomogeneities must be taken into account even if the methods to do so are still in their infancy.

More reasons why approximate results are uncertain, misleading or not fully useful will be given later in this book.

During the 80 years after the Friedmann–Lemaître–Robertson–Walker papers, relativity has advanced much farther on many fronts and produced results that can be directly used in interpreting cosmological observations. In the present text we demonstrate several examples of such applications of exact methods of relativity, mostly (but not exclusively) taken from our own papers. This line of research has a tradition going back to the 1930s, and many powerful exact results had been derived long ago – they are presented systematically in the review by Krasiński (1997). Here we concentrate on the papers that relied on modern astronomical data.

In the first part of this book we present those elements of exact relativistic cosmology that are directly applicable to the interpretation of observations. We assume that the reader is familiar with relativity at the graduate level. We describe the basic properties of three classes of models that have already proved usable: (I) The Lemaître (1933)–Tolman (1934) (L–T) model, which is an exact solution of Einstein's equations with spherical symmetry and dust source – the simplest generalisation of the Friedmann models. (II) The model for which we propose the name 'Lemaître model' – for good historical reasons, see Lemaître (1933) and Sec. 2.2, but which is known in the literature as the Misner–Sharp (1964) approach. This is not an explicit solution, but a metric whose components obey a set of two equations. This scheme, well suited to numerical treatment, describes the evolution of a spherically symmetric perfect fluid with nonzero pressure gradient. (III) The Szekeres (1975) model, which is an exact dust solution with no symmetry, generalising L–T. This is currently the most sophisticated known exact solution of Einstein's equations of cosmological relevance. In Part I of the book we also introduce the relativistic description of light propagation, which includes a detailed presentation of apparent and event horizons in the L–T model. This is a necessary introduction to the discussion of formation and evolution of nonstationary black holes.

In Part II we then present and discuss the applications of exact relativistic methods to actual problems of observational cosmology. These include:

1. **Formation of a galaxy with a central black hole.** This is a problem for whose solution exact methods of relativity are essential, since it is impossible even to define a black hole in a meaningful way by Newtonian methods or perturbations of an FLRW background. Since this was the first investigation of this kind, we used the simplest model that was applicable to it – the L–T model. Its spherical symmetry is a drawback, since most galaxies are not spherically symmetric, and they rotate in addition. However, our investigation may be useful for preliminary qualitative understanding of the process. We considered two possible mechanisms of formation of such an entity: a gravitational collapse of an ordinary ball of dust, and a condensation forming around a pre-existing wormhole.

   We aimed to reproduce the density distribution and mass of one actually observed galaxy with a central black hole, M87. For this, we used the density profile in that galaxy, believed to be known from observations, and for the density profile within the growing black hole, about which nothing is and nothing can be known from observations, we chose a simple model, joined onto the galaxy profile.

   To set realistic initial conditions we assumed that by the time of decoupling of matter and radiation the condensation that would later grow to become the galaxy had a density contrast consistent with the implications of measurements of temperature anisotropies of the CMB radiation. There is a problem with this that keeps coming up in several other investigations: the currently best achieved angular resolution of measurements of temperature is $0.1°$, while a typical proto-galaxy would occupy a region of angular diameter of only $0.004°$ on the CMB sky (see Table 4.1). Thus, at present there are just no adequate observational data to constrain our model. Lacking any better choice, we took the limits known for the scale of $0.1°$ to apply to a single galaxy.

It turned out that the galaxy–black hole structure can be generated by both mechanisms, but for the condensation around a wormhole the black hole appears in a much shorter time, almost instantly after the Big Bang. In that section we stressed that the horizon whose position can be approximately determined from observations is the apparent horizon, not the event horizon.

2. **Formation and evolution of galaxy clusters and voids, using the L–T model.** Inhomogeneities are naturally generated in an inhomogeneous model by the form of the arbitrary functions defining the model. We did not therefore discuss their origin.

Rather, we inferred the amplitude of perturbations at the time of decoupling from the constraints imposed by the measured temperature fluctuations of the CMB radiation. Then we showed that perturbations at decoupling that are consistent with observational constraints can generate a galaxy cluster whose calculated parameters corresponding to the present epoch $(1.5 \times 10^{10}$ y after the Big Bang) agree with the observed parameters of a galaxy cluster chosen at random from the Abell catalogue. Here we encountered again the problem mentioned above: the observational upper limit on the temperature anisotropy of the CMB radiation is determined at angular scales about 20 times larger than the angular diameter that an Abell proto-cluster would occupy on the CMB sky. With better resolutions sure to be achieved in the future, our constraints will have to be re-evaluated. For voids, by the same method, we achieved a qualitative agreement with observations, but for the density within a void we obtained values several times larger than those observed.

The presence of non-baryonic dark matter (which does not interact with photons) would solve this discrepancy, since at the last scattering instant the amplitude of dark matter fluctuations could be larger than temperature fluctuations. However, this discrepancy can also be removed by more conventional methods, which do not require large fluctuations at last scattering. It may perhaps be removed by a more careful choice of the amplitudes and profiles of density and velocity in the proto-voids, but meanwhile one of us (K.B.) put forward another solution of the problem (see below).

These investigations also demonstrated that velocity perturbations are distinctly more efficient in generating structures than are density perturbations. This calls for a revision of the long-standing structure formation paradigm that relies on the belief that density perturbations alone are responsible for the origin of structures. Matter at the edge of today's typical void should occupy a region of angular diameter 0.1° on the CMB sky, which is comparable to the current best resolution achieved in observations. Thus, for voids we are on the verge of being able to test our models against observations.

3. **A more precise description of the formation of voids.** In order to solve the problem of insufficiently low densities within voids obtained in the approach discussed above, we used the Lemaître model describing a mixture of inhomogeneous dust and inhomogeneous radiation. Since the Einstein equations with no further assumptions are indeterminate in this case (the number of unknown functions is larger by one than the number of equations), something had to be assumed about the radiation distribution. Lacking any data, we took the simplest assumption – that the comoving spatial extent of the perturbations of radiation density does not change with time (however the amplitude of radiation density does change with time in the same way as in the Friedmann models). Proceeding from this, we were able to reproduce the current density distribution in voids with arbitrary precision. The assumption about the profile of radiation density will most probably have to be modified in the future, but our result shows that this is a workable approach to the problem of void formation. This investigation highlights the fact that the decoupling of matter and radiation is not a single instant, as is usually assumed for simplicity, but a process extended in time.

4. **The evolution of double structures (cluster–void pairs).** We showed that the evolution of pairs where a void sits on the edge of a cluster, or vice versa, can be described using the quasi-spherical Szekeres (1975) model. Actually, the Szekeres model can be used to describe multiple structures as well, but because of its mathematical limitations the parameters of the additional objects are no longer free; in particular the peripheral structures come out too large. This model helps us to understand some actually observed facts, such as a faster evolution of large voids at larger distances from condensations compared to voids in the proximity of condensations. An additional bonus of this investigation is a physical interpretation of the arbitrary functions in the Szekeres model (they define the direction of the dipole component of the density distribution).

5. **Interpretation of the type Ia supernova dimming as an effect of inhomogeneities.** We showed that, in principle, the observed excessive dimming of type Ia supernovae with distance can be accounted for, on the basis of proper general relativity, without introducing 'dark energy', if one permits inhomogeneity in the matter distribution. This description is more natural than the standard assumption, which postulates a new and completely unknown form of energy. This might also be a way of dealing with the 'coincidence problem' attached to the time when the cosmological constant becomes

dominant over the matter density. Even if the cosmological constant appears as a very natural geometrical tool in Einstein's equations, this implies that we live in a preferred epoch, when the densities of ordinary matter and 'dark energy' are comparable. (The cosmological term becomes negligible at very high mass densities and extremely dominant at very low mass densities.)

We described how this problem can be dealt with in the more general setting, then we gave some examples of exact inhomogeneous models with no cosmological constant which can be found in the recent literature, and which fit cosmological observations such as supernova observations, baryon acoustic oscillations and $H(z)$ measurements, and in addition recover the position of the CMB power spectrum peaks. It was thus proved that evolving inhomogeneities can mimic, partly or totally, the effects of the 'dark energy' component of the Concordance model.

Note, however, that the 'dark energy' problem can be approached in a still different way. The cosmological models of the Robertson–Walker class are supposed to apply at large scales, to an already-averaged geometry and matter distribution in the Universe. So far, there exists no universally accepted definition of averaging that would be both exact and covariant. However, there are several definitions of approximate or non-covariant averaging procedures (see, for example, the last chapter in Krasiński, 1997, and more recent references in Célérier, 2007b) which agree in one point: averaging the nonlinear Einstein equations produces an additional term that mimics negative pressure (i.e. repulsion) in the large-scale energy–momentum tensor. We did not discuss this approach here because of the lack of a general consensus concerning the right method, but this line of research also seems more physical than postulating a new form of energy.

6. **Solving the 'horizon problem' without the use of inflationary models.** We showed that this supposed problem can be solved using the L–T model with appropriately chosen arbitrary functions. This is preferable to the use of inflation, which makes assumptions about physical conditions in the Universe at such early times that any kind of direct verification is currently impossible. Moreover, the solution we proposed solves the horizon problem whatever the location of the observer in spacetime while inflation solves it only temporarily. Alternatives to inflation deserve at least to be considered, but, unfortunately, they tend to be suppressed in the noise of lobbying that has surrounded the inflationary paradigm from the very beginning.

7. **Influence of inhomogeneous structures in the path of a light ray on the observed temperature distribution of the CMB radiation.** We first examined whether (part of) the dipole moment of the CMB might be of a cosmological nature, by putting the observer off the centre of a particular class of L–T models. This result must be taken with caution, since the moments beyond the quadrupole were not calculated here. However, it was

the first time, to our knowledge, that the geodesic equations for non-radial photons and formulae for the dipole and the quadrupole were established for L–T models. We also gave an account of a study similar to ours, which was completed later on and in which the octopole was also calculated. For further investigation we used the quasi-spherical Szekeres model to describe localised condensations or voids, matched into the homogeneous Friedmann background. The light ray under investigation proceeded from a source that emitted it during the decoupling epoch, through several inhomogeneous structures, to an observer registering it at the present epoch. It turned out that the temperature anisotropies caused by the structures are smaller than those generated by the Sachs–Wolfe (1967) effect, unless the observer is situated within one of those structures.

The investigations reported here are preliminary, and most of them will have to be refined or revised in the future, when more precise data become available, and methods of self-consistent interpretation of observational data against inhomogeneous models are developed. The point we wish to make at present is that this branch of cosmology is already advanced to some degree and its results and potential contributions deserve to be more widely recognised.

PART I

Theoretical background

2

Exact solutions of Einstein's equations that are used in cosmology

2.1 The Lemaître–Tolman model

2.1.1 Basic properties

The Lemaître (1933)–Tolman (1934) (L–T) model is a spherically symmetric nonstatic solution of the Einstein equations with a dust source, i.e. the matter tensor is $T^{\alpha\beta} = \rho u^{\alpha} u^{\beta}$. The coordinates are assumed to be comoving, so that the 4-velocity of matter is $u^{\alpha} = \delta^{\alpha}_t$. See Krasiński (1997) and Plebański and Krasiński (2006) for an extensive list of properties and other work on this model. Its metric is (in units in which $c = 1$ and in the synchronous time gauge):

$$\mathrm{d}s^2 = \mathrm{d}t^2 - \frac{R_{,r}{}^2}{1 + 2E(r)}\mathrm{d}r^2 - R^2(t,r)(\mathrm{d}\vartheta^2 + \sin^2\vartheta\,\mathrm{d}\varphi^2), \qquad (2.1)$$

where $E(r)$ is an arbitrary function, $R_{,r} = \partial R/\partial r$, and $R(t,r)$ obeys

$$R_{,t}{}^2 = 2E + \frac{2M}{R} + \frac{\Lambda}{3}R^2, \qquad (2.2)$$

where $R_{,t} = \partial R/\partial t$ and Λ is the cosmological constant. Equation (2.2) is a first integral of the Einstein equations, and $M = M(r)$ is another arbitrary function of integration. The mass density in energy units is:

$$\kappa\rho = \frac{2M_{,r}}{R^2 R_{,r}}, \qquad \text{where } \kappa = \frac{8\pi G}{c^4}. \qquad (2.3)$$

The metric of the space $t = \text{const}$ would be flat if $E(r)$ were set to zero, so E determines the curvature of space at each r value. Comparison of (2.2) with the Newtonian energy equation for a spherically symmetric dust distribution indicates that $M(r)$ is the gravitational mass contained within the comoving spherical shell at any given r, while $E(r)$ is the energy per unit mass of the particles in that shell.

Equation (2.2) can be solved by simple integration:

$$\int_0^R \frac{\mathrm{d}\widetilde{R}}{\sqrt{2E + 2M/\widetilde{R} + \frac{1}{3}\Lambda\widetilde{R}^2}} = t - t_B(r), \qquad (2.4)$$

where t_B appears as an integration function (the bang time), and is an arbitrary function of r. This means that the Big Bang is not simultaneous as in the Friedmann models, but occurs at different times at different distances from the origin.

For the beginning we assume $\Lambda = 0$ (but see Sec. 2.1.7). Then (2.2) can be solved explicitly, and the solutions are: when $E < 0$ (elliptic evolution):

$$R(t,r) = \frac{M}{(-2E)}(1 - \cos\eta), \tag{2.5a}$$

$$\eta - \sin\eta = \frac{(-2E)^{3/2}}{M}(t - t_B(r)), \tag{2.5b}$$

where η is a parameter; when $E = 0$ (parabolic evolution):

$$R(t,r) = \left[\frac{9}{2}M(t - t_B(r))^2\right]^{1/3}; \tag{2.6}$$

and when $E > 0$ (hyperbolic evolution):

$$R(t,r) = \frac{M}{2E}(\cosh\eta - 1), \tag{2.7a}$$

$$\sinh\eta - \eta = \frac{(2E)^{3/2}}{M}(t - t_B(r)). \tag{2.7b}$$

Note that all the formulae given so far are covariant under coordinate transformations of the form $\tilde{r} = g(r)$. This means one of the functions $E(r)$, $M(r)$ and $t_B(r)$ can be fixed at our convenience by the choice of g. We can define a scale radius and a scale time for each worldline with

$$P(r) = \frac{2M}{|2E|}, \qquad T(r) = \frac{2\pi M}{|2E|^{3/2}}, \tag{2.8}$$

and it is evident from (2.5) that, for the elliptic case, these are the maximum R and the lifetime for each r value. The crunch time t_C in the elliptic case is then

$$t_C(r) = t_B(r) + T(r). \tag{2.9}$$

Rewriting (2.5b) or (2.7b) we obtain

$$\eta - \sin\eta = 2\pi(t - t_B)/(t_C - t_B), \tag{2.10a}$$

$$\sinh\eta - \eta = 2\pi(t - t_B)/T, \tag{2.10b}$$

so in elliptic models larger η means that the dust worldline has completed a larger fraction of its lifetime between the bang and the crunch, and in hyperbolic models, η increases with the number of scale times completed.

The parametric solutions (2.5) and (2.7) can also be written

$$t = t_B + \frac{M}{(-2E)^{3/2}}\left\{\arccos\left(1 + \frac{2ER}{M}\right) - 2\sqrt{\frac{-ER}{M}\left(1 + \frac{ER}{M}\right)}\right\},$$

$$0 \le \eta \le \pi, \tag{2.11}$$

$$t = t_B + \frac{M}{(-2E)^{3/2}}\left\{\pi + \arccos\left(-1 - \frac{2ER}{M}\right)\right.$$

$$\left. + 2\sqrt{\frac{-ER}{M}\left(1 + \frac{ER}{M}\right)}\right\}, \quad \pi \le \eta \le 2\pi, \tag{2.12}$$

for the expanding and collapsing elliptic cases, and

$$t = t_B + \frac{M}{(2E)^{3/2}} \left\{ 2\sqrt{\frac{ER}{M}\left(1 + \frac{ER}{M}\right)} - \text{arcosh}\left(1 + \frac{2ER}{M}\right) \right\} \quad (2.13)$$

for the hyperbolic case (expanding).

Apart from extended parabolic regions, there are also parabolic boundaries between elliptic and hyperbolic regions, where $E \to 0$, but $E_{,r} \neq 0$. The limiting forms of equations (2.5) and (2.7) are found by requiring well-behaved time evolution and setting $\eta = \tilde{\eta}\sqrt{E}$, so that $\tilde{\eta}$ is finite if $(t - t_B)$ is (see Hellaby and Lake, 1985; Hellaby and Krasiński, 2006).

The Friedmann models are contained in the L–T class as the limit:

$$t_B = \text{const}, \qquad |E|^{3/2}/M = \text{const}, \quad (2.14)$$

and one of the standard radial coordinates for the Friedmann model results if the coordinates in (2.5)–(2.7) are additionally chosen so that:

$$M = M_0 r^3 \quad \to \quad E = E_0 r^2 \quad (2.15)$$

with M_0 and E_0 being constants.

The Schwarzschild solution in the Lemaître (1933)–Novikov (1964) coordinates results as the case $M = \text{constant}$ and $\Lambda = 0$. The other functions, $E(r)$ and $t_B(r)$, may be arbitrarily specified, and this must give at least part of the maximally extended Kruskal (1960)–Szekeres (1960) manifold. Whether the manifold is fully covered or not depends on the forms of the functions $E(r)$ and $t_B(r)$ (see Hellaby 1987, 1996a).

2.1.2 Origin conditions

An origin, or centre of spherical symmetry, occurs at $r = r_c$ where $R(t, r_c) = 0$ for all t. The conditions for a regular centre follow from the requirement that there is no point-mass and no curvature singularity at $r = r_c$ at all times after the Big Bang and before the Big Crunch. The limits $r \to r_c$ below are taken along a constant-t hypersurface.

The requirement of no point-mass implies $M(r_c) = 0$. The components of the Riemann tensor in the orthonormal tetrad implied by the metric (2.1)–(2.2) (with coordinates labeled as $(t, r, \vartheta, \varphi) = (x^0, x^1, x^2, x^3)$) are

$$R_{0101} = \frac{1}{3}\Lambda + \frac{2M}{R^3} - \frac{M_{,r}}{R^2 R_{,r}}, \qquad R_{0202} = R_{0303} = \frac{1}{3}\Lambda - \frac{M}{R^3},$$

$$R_{1212} = R_{1313} = -\frac{1}{3}\Lambda + \frac{M}{R^3} - \frac{M_{,r}}{R^2 R_{,r}}, \qquad R_{2323} = -\frac{1}{3}\Lambda - \frac{2M}{R^3}.$$

These are all scalars, and it can be seen that with $M(r_c) = R(t, r_c) = 0$ all we need to make the curvature nonsingular at $r = r_c$ is a finite limit of M/R^3 at

$r = r_c$. We have, using (2.3)

$$\lim_{r \to r_c} \frac{M}{R^3} = \lim_{r \to r_c} \frac{M_{,r}}{3R^2 R_{,r}} = \frac{1}{6}\kappa\rho(t, r_c),$$

thus, the curvature is finite at $r = r_c$ if and only if $\rho(t, r_c)$ is. The above implies

$$\lim_{r \to r_c} \frac{R}{M^{1/3}} = \left(\frac{6}{\kappa\rho(t, r_c)}\right)^{1/3},$$

which is finite when $\rho(t, r_c) \neq 0$. The behaviour of E as $r \to r_c$ is then determined by (2.2), which implies (2.16c) below.

The conditions discussed above are sufficient to guarantee that all scalar components of curvature are finite at $r = r_c$. However, it is often required in addition that $\rho_{,\overline{R}} \xrightarrow[r \to r_c]{} 0$, where $\overline{R}$ is R taken at any constant time. This requirement implies (2.16d) and (2.16e) below.

In the equations below, the symbol $\mathcal{O}_d(M)$ denotes a function that has the property $\lim_{M \to 0}\left(\mathcal{O}_d(M)/M^d\right) = 0$. The resulting conditions for the neighbourhood of r_c are ($\beta(t)$ is an arbitrary function and γ, τ_0 and τ are arbitrary constants):

$$M(r_c) = 0, \tag{2.16a}$$

$$R = \beta(t)M^{1/3} + \mathcal{O}_{1/3}(M) \quad \text{along constant } t, \quad \beta \neq 0 \tag{2.16b}$$

$$E = \gamma M^{2/3} + \mathcal{O}_{2/3}(M), \tag{2.16c}$$

$$t_B = \tau_0 + \tau\mathcal{O}_{1/3}(M), \tag{2.16d}$$

$$\kappa\rho = 6/\beta^3 + \mathcal{O}_{1/3}(M) \quad \text{along constant } t. \tag{2.16e}$$

We also need $\tau < 0$ to avoid shell crossings. Without the extra demand that $\rho_{,\overline{R}} \xrightarrow[r \to r_c]{} 0$, in the last two equations above we would have $\mathcal{O}_0(M)$.

The L–T models that do not obey this extra condition are sometimes said to have a 'weak singularity' at the centre, because then $\Box\mathcal{R} \to \infty$ at $r = r_c$, where $\Box$ is the covariant d'Alembertian, and $\mathcal{R}$ is the Ricci scalar (see Vanderveld *et al.*, 2006). This is, however, a misnomer. A singularity may be weak when it is there, but when there is no singularity at all, we cannot speak about a 'weak singulariy' (Krasiński *et al.*, 2009).

The extra condition is equivalent to $R_{,rr}|_{r=0} = 0$, where $r \propto M^{1/3}$ is the radial comoving coordinate. It can be checked from Eq. (19) of Mustapha and Hellaby (2001) that this requirement is always satisfied if Eqs. (2.16) hold. See also Humphreys, Maartens and Matravers (1998).

2.1.3 Shell crossings, maxima and minima

The density and the Kretschmann scalar diverge where $R_{,r} = 0$ and $M_{,r} \neq 0$, but loci where $M_{,r}/R_{,r}$ is finite when $R_{,r} = 0$ are also possible on constant r

shells, and these constitute regular extrema in the spatial section of constant t. Both $M_{,r}$ and $R_{,r}$ change sign across a regular extremum. A closed model must necessarily have two origins and a spatial maximum (a 'belly'). A vacuum model must necessarily have a spatial minimum (a 'neck' or 'wormhole'). Necks may exist also in nonvacuum models, see Sec. 4.2 later in this text.

Shell crossings, where a constant r shell collides with its neighbour, are loci of $R_{,r} = 0$ that are not regular maxima or minima of R. They create undesirable singularities where the density diverges and changes sign. The conditions on the 3 arbitrary functions that ensure none be present anywhere in an L–T model, as well as those for regular maxima and minima in spatial sections, were given by Hellaby and Lake (1985), and will be used below.

The condition $E \geq -1/2$ must be fulfilled in any physical L–T spacetime, as seen from (2.1); only then is the signature $(+ - --)$. The equality $E = -1/2$ must occur at regular maxima and minima (see Hellaby and Lake, 1985; Plebański and Krasiński, 2006), but these locations are either inside 'wormholes', or at very large cosmological distances from the origin.

2.1.4 L–T models of cosmic structures

In later chapters we present L–T models of structures such as galaxy clusters and voids, as well as the supernova observations.

Now, the L–T model has two disadvantages. Because of its spherical symmetry, we cannot take rotation into account. In consequence, the model does not apply to single galaxies, and is a rough approximation to galaxy clusters. The objects to which the model can be best applied are voids because for them it has been proved (by Newtonian methods) that spherical symmetry is a stable property (Sato and Maeda, 1983). Because of zero pressure, the model is unable to describe any hydro- or thermodynamics, and so is not applicable to the early Universe and to the late stages of the collapse of matter.

Among the advantages, our method yields structures reasonably similar to galaxy clusters and galaxies with central black holes from initial conditions that are in agreement with the CMB observations, and highlights the interplay between the density and velocity fluctuations. Using exact solutions of Einstein's equations, we can track the evolution of the structures up to the present and into the future, linear or nonlinear 'regimes' being irrelevant, since in an exact model there are no 'regimes' at all.

The choice of such simple inhomogeneous models to represent large-scale structures in the Universe can, of course, be questioned. The assumptions of spherical symmetry and a central observer are generally made for simplicity, but they are grounded on the observed quasi-isotropy of the CMB temperature. They are less restricted than the Friedmann models, having relaxed one degree of symmetry relative to the homogeneity assumption, and they provide a reasonable working approximation in which the Universe is averaged only over the angular coordinates around the observer, but with no radial smoothing.

However, some models assuming an off-centre observer have also been studied, to avoid possible misleading features of spherical symmetry. They will be presented in Secs. 3.1.2, 3.1.3, 7.2 and 7.3.

Other methods of dealing with some of the disadvantages will be presented in the following sections. For example, in the Szekeres (1975) model one can consider more general shapes of the galaxy clusters and voids (see Sec. 4.6). One can also solve numerically the Einstein equations for a spherically symmetric distribution of dust mixed with inhomogeneous radiation, and in this way partly overcome the disadvantage of zero pressure. The results reproduce the real density profiles of voids rather well (see Sec. 4.5).

2.1.5 Defining a model by initial and final profiles with $\Lambda = 0$

1. A reformulation of the evolution equations

In the traditional approach to dynamics, one formulates the initial data (e.g. positions and velocities of particles) at a time t_1, then one uses the field equations to evolve the data to $t_2 > t_1$, and compares the calculated results at t_2 with observations or experiments. However, in cosmology the initial state at t_1 (taken, for example, to be at last scattering) is accessible to observations only indirectly and with a rather imperfect precision. The observational data are collected at the final instant t_2, and only some of them can be projected back to t_1. Using the traditional approach, one assumes various data at t_1 and tries to 'shoot' into the observed final state at t_2.

The evolution equations can be reformulated so that information from both t_1 and t_2 can be fed into the models. We present here such an approach to the L–T model (Krasiński and Hellaby, 2002 and 2004a). We give numerical examples of evolution of small initial density or velocity fluctuations at t_1 (= last scattering epoch) to modern structures such as galaxy clusters and voids at $t_2 = $ now. The reformulated equations have unique solutions under conditions that are easy to verify.

In Chapter 4 we will use the L–T model to connect an initial state of the Universe, defined by a mass-density or velocity distribution, to a final state defined also by one of these distributions. We will give numerical examples in which we will incorporate the actual astronomical data on the distribution of temperature of the background radiation and on the mass distribution in the currently existing clusters, voids and galaxies with central black holes. In Chapter 6, the L–T model will be used to interpret the observational data on the type Ia supernovae and to deal with the 'horizon problem'.

In constructing our models of galaxy clusters and voids, it will be convenient to use $M(r)$ as the radial coordinate (i.e. $\tilde{r} = M(r)$) – because in most sections we shall not need to pass through any 'necks' or 'bellies'. Thus, $M(r)$ will be a strictly growing function in the whole region under consideration. In some of the sections we shall consider a black hole with a 'neck' or 'wormhole', but even there we will consider only one side of the wormhole, where $M(r)$ is also increasing. Then with $R = R(t, M)$:

$$\kappa\rho = \frac{2}{R^2 \partial R/\partial M} = \frac{6}{\partial(R^3)/\partial M}, \tag{2.17}$$

and this is inverted so that $R(t, r)$ is defined by a given $\rho(t, r)$:

$$R^3 = \int_0^M \frac{3c^4}{4\pi G\rho(t, x)}\,\mathrm{d}x \tag{2.18}$$

(the last equation results with the initial condition $R = 0$ at $M = 0$ which we assume in nearly all cases).

For clarity of the calculations, we introduce the following quantities:

$$a = R/M^{1/3}, \tag{2.19a}$$

$$x = |2E|/M^{2/3}, \tag{2.19b}$$

$$b = R_{,t}/M^{1/3}. \tag{2.19c}$$

2. **Part I: density to density evolution**

We specify the density distributions at the instants $t = t_i$, $i = 1, 2$:

$$\rho(t_i, M) = \rho_i(M), \tag{2.20}$$

and calculate the corresponding $R(t_i, M)$ by (2.18). Throughout this text we assume $R_{,t}(t_1, M) > 0$, except in a region describing a black hole. This assumption is dictated by the intended application of our results (structure formation in the Universe), but a similar investigation could be done for collapsing matter. For definiteness, we also assume $t_2 > t_1$ throughout.

Hyperbolic evolution, $E > 0$

We write out the evolution equations (2.7) at t_1 and t_2:

$$R_i(M) = R(t_i, M) = \frac{M}{2E}(\cosh\eta_i - 1),$$

$$\sinh\eta_i - \eta_i = \frac{(2E)^{(3/2)}}{M}[t_i - t_B(M)], \tag{2.21}$$

and $R_2(M) > R_1(M)$ in consequence of $\rho(t_2, M) < \rho(t_1, M)$. Solving for t_B at t_1 we get

$$t_B = t_1 - \frac{M}{(2E)^{3/2}}\left[\sqrt{(1 + 2ER_1/M)^2 - 1} - \text{arcosh}(1 + 2ER_1/M)\right]. \tag{2.22}$$

We substitute this in (2.21) at t_2, obtaining

$$\sqrt{(1 + 2ER_2/M)^2 - 1} - \text{arcosh}(1 + 2ER_2/M)$$
$$- \sqrt{(1 + 2ER_1/M)^2 - 1} + \text{arcosh}(1 + 2ER_1/M)$$
$$= [(2E)^{3/2}/M](t_2 - t_1). \tag{2.23}$$

This equation defines $E(M)$, and then (2.22) defines $t_B(M)$. These two functions completely specify the L–T evolution from $\rho(t_1, M)$ to $\rho(t_2, M)$. For ease of calculations, let us denote (see (2.19)):

$$\psi_H(x) \stackrel{\text{def}}{=} \sqrt{(1 + a_2 x)^2 - 1} - \operatorname{arcosh}(1 + a_2 x)$$
$$- \sqrt{(1 + a_1 x)^2 - 1} + \operatorname{arcosh}(1 + a_1 x) - (t_2 - t_1) x^{3/2}. \tag{2.24}$$

Now we have to answer the question: for what values of $a_2 > a_1$ and $t_2 > t_1$, does the equation $\psi_H(x) = 0$ have a solution $x \neq 0$? An elementary reasoning (Krasiński and Hellaby, 2002) leads to the conclusion that the equation has a solution if and only if the following inequality is fulfilled:

$$t_2 - t_1 < \frac{\sqrt{2}}{3} \left(a_2{}^{3/2} - a_1{}^{3/2} \right), \tag{2.25}$$

and then the solution is unique. This inequality says that the expansion between t_1 and t_2 must have been faster than it would be in the $E = 0$ model.

The result above only shows the existence of a solution for a given value of M. Some initial conditions may lead to shell crossings, and these are not excluded by (2.25). But the criteria for the occurrence of shell crossings are well investigated (Hellaby and Lake, 1985). If shell crossings occur outside the range where the model is supposed to apply (e.g. before t_1), then they are not a problem – the L–T model cannot describe those epochs anyway.

Elliptic evolution, $E < 0$

The evolution must be considered separately for the final state expanding [for which $\eta \in [0, \pi]$ in (2.5)] and for the final state recollapsing [$\eta \in [\pi, 2\pi]$ in (2.5)].

For the still-expanding final state, the ψ-function is:

$$\psi_X(x) \stackrel{\text{def}}{=} \arccos(1 - a_2 x) - \sqrt{1 - (1 - a_2 x)^2} - \arccos(1 - a_1 x)$$
$$+ \sqrt{1 - (1 - a_1 x)^2} - (t_2 - t_1) x^{3/2}, \tag{2.26}$$

the definitions of a_i and x being (2.19a) and (2.19b).

The reasoning is analogous to the one for (2.24), but the arguments of arccos must have absolute values not greater than 1. This implies $0 \le x \le 2/a_i$ for both i, and so, since $a_2 > a_1$,

$$0 \le x \le 2/a_2, \tag{2.27}$$

which means: if there is any solution of $\psi_X(x) = 0$, then it will have the property (2.27). Equation (2.27) is equivalent to $(R_{,t})^2 \ge 0$ at both t_1 and t_2.

Again, an elementary reasoning (Krasiński and Hellaby, 2002) leads to the conclusion that $\psi_X(x) = 0$ has a solution if and only if two inequalities are fulfilled at the same time:

$$\frac{\sqrt{2}}{3}\left(a_2{}^{3/2} - a_1{}^{3/2}\right) < t_2 - t_1$$

$$\leq (a_2/2)^{3/2}\left[\pi - \arccos(1 - 2a_1/a_2) + 2\sqrt{a_1/a_2 - (a_1/a_2)^2}\right]. \quad (2.28)$$

With (2.28) fulfilled, the solution is unique. The first inequality means that the model must have expanded between t_1 and t_2 slower than the $E = 0$ model would have; the second inequality means that the final state is still earlier than the instant of maximal expansion.

As is seen from (2.1), the function E must obey $E \geq -1/2$, or else the signature of the spacetime will become unphysical. This condition does not follow from (2.28) and has to be verified separately, together with the no-shell-crossing conditions.[1]

For the recollapsing final state, x is defined as in (2.19b), but the ψ-function is:

$$\psi_C(x) \stackrel{\text{def}}{=} \pi + \arccos(-1 + a_2 x) + \sqrt{1 - (1 - a_2 x)^2}$$

$$- \arccos(1 - a_1 x) + \sqrt{1 - (1 - a_1 x)^2} - (t_2 - t_1)x^{3/2}. \quad (2.29)$$

The solution of $\psi_C(x) = 0$ exists if and only if

$$t_2 - t_1 \geq (a_2/2)^{3/2}\left[\pi - \arccos(1 - 2a_1/a_2) + 2\sqrt{a_1/a_2 - (a_1/a_2)^2}\right], \quad (2.30)$$

and then the solution is unique (see again Krasiński and Hellaby (2002) for details).

3. **Part II: velocity to density evolution**

A useful measure of velocity is $b_i(M) \stackrel{\text{def}}{=} R_{,t}(t, M)/M^{1/3}\big|_{t=t_i}$; its variations indicate inhomogeneity in the expansion rate, since it becomes constant in the Friedmann limit. Suppose that the initial state is specified by a velocity profile $b_1(M)$, while the final state is specified by a density distribution $\rho(t_2, M)$. The scheme of the previous subsection goes through in a very similar way.

Hyperbolic evolution, $E > 0$

With the variables defined as in (2.19a) and (2.19c), the equation to be solved is $\Phi_H(x) = 0$, where

$$\Phi_H(x) \stackrel{\text{def}}{=} \sqrt{(1 + a_2 x)^2 - 1} - \sqrt{\left(\frac{b_1{}^2 + x}{b_1{}^2 - x}\right)^2 - 1}$$

$$- \text{arcosh}(1 + a_2 x) + \text{arcosh}\left(\frac{b_1{}^2 + x}{b_1{}^2 - x}\right) - x^{3/2}(t_2 - t_1). \quad (2.31)$$

---

[1] But, as noted in Krasiński and Hellaby (2002), it is easy to extend through a spatial extremum in R and M, where $E = -1/2$, should one occur. Past that point, both $R_{,r}$ and $M_{,r}$ change sign and E increases again.

Now the necessary and sufficient condition for the solvability of $\Phi_H(x) = 0$ is:

$$t_2 - t_1 < \frac{\sqrt{2}}{3} a_2{}^{3/2} - \frac{4}{3b_1{}^3}, \qquad (2.32a)$$

$$2/a_2 < b_1{}^2. \qquad (2.32b)$$

The first of these is equivalent to (2.25) (Krasiński and Hellaby, 2004a). The second one is a necessary condition for the existence of a $t_2 > t_1$ obeying (2.32a), and is equivalent to $R_2 > R_1$. With both inequalities fulfilled, the solution of $\Phi_H(x) = 0$ is unique.

Elliptic evolution, $E < 0$

When the final state is still expanding, we must find the root of Φ_X, where

$$\Phi_X(x) \overset{\text{def}}{=} \sqrt{1 - \left(\frac{b_1{}^2 - x}{b_1{}^2 + x}\right)^2} - \sqrt{1 - (1 - a_2 x)^2}$$

$$+ \arccos(1 - a_2 x) - \arccos\left(\frac{b_1{}^2 - x}{b_1{}^2 + x}\right) - x^{3/2}(t_2 - t_1), \qquad (2.33)$$

and the necessary and sufficient condition for the existence of an L–T evolution between the two given states is:

$$\frac{\sqrt{2}}{3} a_2{}^{3/2} - \frac{4}{3b_1{}^3} < t_2 - t_1 \leq (a_2/2)^{3/2}\left[\pi + \frac{b_1\sqrt{2a_2}}{a_2 b_1{}^2/2 + 1}\right.$$

$$\left. - \arccos\left(\frac{a_2 b_1{}^2/2 - 1}{a_2 b_1{}^2/2 + 1}\right)\right]. \qquad (2.34)$$

This set of inequalities is equivalent to (2.28). With the inequalities fulfilled, the solution of the corresponding equation is unique (see Krasiński and Hellaby, 2004a).

When the final state is recollapsing, the $E < 0$ evolution exists if and only if

$$t_2 - t_1 \geq (a_2/2)^{3/2}\left[\pi + \frac{b_1\sqrt{2a_2}}{a_2 b_1{}^2/2 + 1} - \arccos\left(\frac{a_2 b_1{}^2/2 - 1}{a_2 b_1{}^2/2 + 1}\right)\right], \qquad (2.35)$$

and x is the solution of $\Phi_C = 0$ where

$$\Phi_C(x) \overset{\text{def}}{=} \sqrt{1 - \left(\frac{b_1{}^2 - x}{b_1{}^2 + x}\right)^2} + \sqrt{1 - (1 - a_2 x)^2}$$

$$+ \pi - \arccos\left(\frac{b_1{}^2 - x}{b_1{}^2 + x}\right) + \arccos(-1 + a_2 x) - x^{3/2}(t_2 - t_1). \qquad (2.36)$$

4. **Part III: velocity to velocity evolution**

The given quantities are now an initial and a final velocity profile $b_1(M)$ and $b_2(M)$. We assume $b_1 > 0$ and $t_2 > t_1$.

Hyperbolic evolution, $E > 0$

Then, in this case we must solve $\chi_H(x) = 0$, where

$$
\chi_H(x) \overset{\text{def}}{=} \sqrt{\left(\frac{b_2{}^2 + x}{b_2{}^2 - x}\right)^2 - 1} - \sqrt{\left(\frac{b_1{}^2 + x}{b_1{}^2 - x}\right)^2 - 1}
$$
$$
- \operatorname{arcosh}\left(\frac{b_2{}^2 + x}{b_2{}^2 - x}\right) + \operatorname{arcosh}\left(\frac{b_1{}^2 + x}{b_1{}^2 - x}\right) - x^{3/2}(t_2 - t_1), \qquad (2.37)
$$

and the necessary and sufficient condition for the existence of an $E > 0$ evolution between the two states is:

$$
0 < b_2 < b_1, \qquad t_2 - t_1 > \frac{4}{3}\left(\frac{1}{b_2{}^3} - \frac{1}{b_1{}^3}\right). \qquad (2.38)
$$

The second inequality becomes clearer when it is rewritten as

$$
b_2{}^3 > \frac{b_1{}^3}{1 + \frac{3}{4}b_1{}^3(t_2 - t_1)}, \qquad (2.39)
$$

which means that an $E > 0$ evolution between the two states will exist provided the velocity of expansion at t_2 is greater than the velocity of expansion of the $E = 0$ model.

Elliptic evolution, $E < 0$

When the final state is still expanding, x is found by solving $\chi_X = 0$, where

$$
\chi_X(x) \overset{\text{def}}{=} -\sqrt{1 - \left(\frac{b_2{}^2 - x}{b_2{}^2 + x}\right)^2} + \sqrt{1 - \left(\frac{b_1{}^2 - x}{b_1{}^2 + x}\right)^2}
$$
$$
+ \arccos\left(\frac{b_2{}^2 - x}{b_2{}^2 + x}\right) - \arccos\left(\frac{b_1{}^2 - x}{b_1{}^2 + x}\right) - x^{3/2}(t_2 - t_1). \qquad (2.40)
$$

The necessary and sufficient condition for the existence of this evolution is the set consisting of the first inequality in (2.38) and

$$
t_2 - t_1 < \frac{4}{3}\left(\frac{1}{b_2{}^3} - \frac{1}{b_1{}^3}\right), \qquad (2.41)
$$

which now implies that the expansion between t_1 and t_2 must have been slower than it would be in an $E = 0$ model.

When the final state is already recollapsing, we have $b_2 < 0$. Then the evolution exists for

$$
b_2 < 0 < b_1 \qquad (2.42)
$$

and x is the root of

$$
\chi_C(x) \overset{\text{def}}{=} \sqrt{1 - \left(\frac{b_1{}^2 - x}{b_1{}^2 + x}\right)^2} - \sqrt{1 - \left(\frac{x - b_2{}^2}{x + b_2{}^2}\right)^2}
$$
$$
+ \pi + \arccos\left(\frac{x - b_2{}^2}{x + b_2{}^2}\right) - \arccos\left(\frac{b_1{}^2 - x}{b_1{}^2 + x}\right) - x^{3/2}(t_2 - t_1). \qquad (2.43)
$$

5. **Borderlines between cases**

Although the above cases cover all other eventualities in theory, the expressions for ψ_H, ψ_X, Φ_H, etc. are not practical numerically near the borderlines between them, since very large terms are subtracted, creating excessive error. The series expansions that apply in the vicinity of the two borderlines – where the evolution is nearly parabolic, or nearly at maximum expansion at t_2 – are given in Krasiński and Hellaby (2002) and Hellaby and Krasiński (2006).

If density or velocity profiles are chosen to match observations or to construct realistic models, the case of an extended parabolic region is so unlikely that it has not been considered here.

2.1.6 Additional methods of specifying models

The initial and final states of an L–T model can be specified in more ways. We explored those in Hellaby and Krasiński (2006); they are as follows:

- a density profile $\rho_i(M)$ is given at time t_i,
- a velocity profile $(R_{,t})_i(M)$ is given at time t_i,
- the bang time is simultaneous,
- the crunch time is simultaneous,
- the time of maximum expansion is simultaneous,
- the model becomes homogeneous at late times,
- only growing modes are present,
- only decaying modes are present,
- a velocity profile $(R_{,t})(M)$ is given at late times,
- a time-scaled density profile $t^3\rho(M)$ is given at late times.

It requires two of the above conditions to specify an L–T model, and in various contexts different combinations may be useful (all combinations of the first two were discussed above). We work them out here in a systematic way for future reference. Where two conditions are specified on the same time slice, this often requires special treatment.

The solutions of (2.5) and (2.7) can be written as follows (Hellaby and Krasiński, 2006):

HX $(E > 0,\ t > t_B)$:
$$t = t_B + x^{-3/2} \left\{ \sqrt{(1 + xa)^2 - 1} - \operatorname{arcosh}(1 + xa) \right\}; \qquad (2.44)$$

EX $(E < 0,\ 0 \leq \eta \leq \pi)$:
$$t = t_B + x^{-3/2} \left\{ \arccos(1 + xa) - \sqrt{1 - (1 + xa)^2} \right\}; \qquad (2.45)$$

EC $(E < 0,\ \pi \leq \eta \leq 2\pi)$:
$$t = t_B + x^{-3/2} \left\{ 2\pi - \arccos(1 + xa) + \sqrt{1 - (1 + xa)^2} \right\}; \qquad (2.46)$$

HC $(E > 0, \ t < t_C)$:

$$t = t_C - x^{-3/2} \left\{ \sqrt{(1 + xa)^2 - 1} - \text{arcosh}(1 + xa) \right\}.$$ (2.47)

HX: $\quad t = t_B + x^{-3/2} \left\{ \sqrt{\left(\dfrac{b^2 + x}{b^2 - x}\right)^2 - 1} - \text{arcosh}\left(\dfrac{b^2 + x}{b^2 - x}\right) \right\};$ (2.48)

EX: $\quad t = t_B + x^{-3/2} \left\{ \arccos\left(\dfrac{b^2 + x}{b^2 - x}\right) - \sqrt{1 - \left(\dfrac{b^2 + x}{b^2 - x}\right)^2} \right\};$ (2.49)

EC: $\quad t = t_B + x^{-3/2} \left\{ 2\pi - \arccos\left(\dfrac{b^2 + x}{b^2 - x}\right) + \sqrt{1 - \left(\dfrac{b^2 + x}{b^2 - x}\right)^2} \right\};$ (2.50)

HC: $\quad t = t_C - x^{-3/2} \left\{ \sqrt{\left(\dfrac{b^2 + x}{b^2 - x}\right)^2 - 1} - \text{arcosh}\left(\dfrac{b^2 + x}{b^2 - x}\right) \right\}.$ (2.51)

1. **Models with a simultaneous bang time**

The condition that the bang is simultaneous,

$$t_{B,r} = 0, \qquad (t_B = \text{specified constant}),$$ (2.52)

is known to generate only growing modes, but as discussed below, does not quite cover all possibilities. We now combine this with a specified density or velocity profile on a given time slice.

Density profile given at time t_i

We specify a density profile $\rho_i(M)$ at time t_i, so the function $a_i(M)$ is determined via (2.18) and (2.19a), and the equations to be solved are (2.45), (2.46) and (2.44).

For the various possible evolution types we find:

HX: $\quad 0 = \sqrt{(1 + a_i x)^2 - 1} - \text{arcosh}(1 + a_i x)$

$$- x^{3/2}(t_i - t_B) \overset{\text{def}}{=} \psi_{BDH}(x);$$ (2.53)

EX: $\quad 0 = \arccos(1 - a_i x) - \sqrt{1 - (1 - a_i x)^2}$

$$- x^{3/2}(t_i - t_B) \overset{\text{def}}{=} \psi_{BDX}(x);$$ (2.54)

EC: $\quad 0 = 2\pi - \arccos(1 - a_i x) + \sqrt{1 - (1 - a_i x)^2}$

$$- x^{3/2}(t_i - t_B) \overset{\text{def}}{=} \psi_{BDC}(x).$$ (2.55)

Equations (2.53)–(2.55) may be solved numerically for $x(M)$ and hence $E(M)$. The conditions for each solution to exist are:

$$\text{HX:} \quad t_i - t_B < \sqrt{2}\, \frac{a_i^{3/2}}{3}, \tag{2.56a}$$

$$\text{EX:} \quad \sqrt{2}\, \frac{a_i^{3/2}}{3} < t_i - t_B < \pi \left(\frac{a_i}{2}\right)^{3/2}, \tag{2.56b}$$

$$\text{EC:} \quad \pi \left(\frac{a_i}{2}\right)^{3/2} < t_i - t_B. \tag{2.56c}$$

In addition, we give series expansions for solutions that are near to an expanding parabolic model (nPX), and near to maximum expansion in an elliptic model (nEM) at t_i, because the above solutions would encounter numerical difficulties close to the parabolic and maximum expansion borderlines:

$$\text{nPX:} \quad x \approx \frac{20}{3a_i}\left(-\frac{\mathrm{d}\tau_P}{\tau_P} + \frac{25\mathrm{d}\tau_P^2}{14\tau_P^2}\right);$$

$$\text{nEM:} \quad x \approx \frac{2}{a_i} - \frac{\mathrm{d}\tau_N^2}{a_i^4} + \frac{3\pi\mathrm{d}\tau_N^3}{2^{5/2}a_i^{11/2}};$$

where

$$\left.\begin{aligned} \tau_i &= t_i - t_B, \\ \tau_P &= \frac{\sqrt{2}}{3}a_i^{3/2}, \qquad \mathrm{d}\tau_P = \tau_i - \tau_P, \\ \tau_N &= \pi\left(\frac{a_i}{2}\right)^{3/2}, \qquad \mathrm{d}\tau_N = \tau_i - \tau_N. \end{aligned}\right\} \tag{2.57}$$

One may think of τ_P as the time it would take a parabolic model to reach a_i, and τ_N as the time to maximum expansion <u>if</u> a_i were the maximum a value. But note that $\mathrm{d}\tau_N$ is not the time since maximum expansion, because when $\mathrm{d}\tau_N = 0$, the model is not exactly at maximum expansion, so a_i and τ_N are less than their maximum values. Note too that $x = 2M/E^{2/3}$ rather than (2.19b) is used to define x for the near parabolic series, so a negative x indicates a slightly elliptic model.

It is important to know which solution to use along each worldline, i.e. for each M value. This is readily apparent from the existence conditions, which are summarised for all pairs of model specifications in Table 2.1.

Velocity profile given at time t_i

For this scenario, we specify $(R_{,t})_i(M)$ at time t_i, giving $b_i(M)$ from (2.19c), and the equations to be solved are (2.49), (2.50) and (2.48), leading to:

$$\text{HX:} \quad 0 = \sqrt{\left(\frac{b_i^2 + x}{b_i^2 - x}\right)^2 - 1} - \operatorname{arcosh}\left(\frac{b_i^2 + x}{b_i^2 - x}\right)$$
$$- x^{3/2}(t_i - t_B) \overset{\text{def}}{=} \psi_{BVH}(x); \tag{2.58}$$

$$\text{EX:} \quad 0 = + \arccos\left(\frac{b_i^2 - x}{b_i^2 + x}\right) - \sqrt{1 - \left(\frac{b_i^2 - x}{b_i^2 + x}\right)^2}$$
$$- x^{3/2}(t_i - t_B) \overset{\text{def}}{=} \psi_{BVX}(x); \tag{2.59}$$

$$\text{EC:} \quad 0 = 2\pi - \arccos\left(\frac{b_i^2 - x}{b_i^2 + x}\right) + \sqrt{1 - \left(\frac{b_i^2 - x}{b_i^2 + x}\right)^2}$$
$$- x^{3/2}(t_i - t_B) \overset{\text{def}}{=} \psi_{BVC}(x). \tag{2.60}$$

The existence conditions for each particular solution are:

$$\text{HX:} \quad b_i > 0 \quad \text{and} \quad t_i - t_B > \frac{4}{3}b_i^3, \tag{2.61a}$$

$$\text{EX:} \quad b_i > 0 \quad \text{and} \quad t_i - t_B < \frac{4}{3}b_i^3, \tag{2.61b}$$

$$\text{EC:} \quad b_i < 0, \tag{2.61c}$$

and the series expansions near the borderlines are:

$$\text{nPX:} \quad x \approx \frac{5b_i^2}{6}\left(\frac{\mathrm{d}\tau_P}{\tau_P} - \frac{25\mathrm{d}\tau_P^2}{28\tau_P^2}\right);$$

$$\text{nEM:} \quad x \approx x_N\left(1 + \frac{16}{3\pi^2}\frac{b_i}{\bar{b}} + \frac{64}{3\pi^4}\frac{b_i^2}{\bar{b}^2}\right);$$

where

$$\left.\begin{array}{ll}\tau_i = t_i - t_B,\\[2mm]\tau_P = \dfrac{4}{3b_i^3}, & \mathrm{d}\tau_P = \tau_i - \tau_P,\\[4mm]x_N = \left(\dfrac{\pi}{\tau_i}\right)^{2/3}, & \bar{b} = \dfrac{2\sqrt{x_N}}{\pi}.\end{array}\right\} \tag{2.62}$$

Here $\bar{b}$ is a kind of average velocity; if the model were exactly at maximum expansion after τ_i, then a would be $a_N = 2/x_N$ and $\bar{b}$ would be a_N/τ_i.

2. **Models that become homogeneous at late times**
We next find those L–T models that approach an R–W model at late times, $t \to \infty$. Only expanding hyperbolic (and parabolic) models have an infinite future, and for these we require

$$2E = K\,M^{2/3}\,, \qquad K = \text{specified constant}. \tag{2.63}$$

Clearly, this case has only decaying modes. The time reverse of this case – the collapsing hyperbolic model – also satisfies the same condition, but was R–W like in the infinite past. See Krasiński (1997) for examples. One may also apply this condition to elliptic models, but the only such model completely free of shell crossings is the R–W model, so we don't consider it here, though part of the evolution of non-homogeneous elliptic models obeying (2.63) will be free of shell crossings and may be of interest.

Density profile given at time t_i

Applying (2.63) to (2.44) for known $a_i(M)$ and t_i leads to

$$\text{HX:} \quad t_B = t_i - K^{-3/2}\left[\sqrt{(1 + a_i K)^2 - 1} - \text{arcosh}(1 + a_i K)\right], \qquad (2.64)$$

which exists if the following relations hold

$$\text{HX:} \quad K > 0 \quad \text{and} \quad t_i - t_B < \sqrt{2}\,\frac{a_i^{3/2}}{3}, \qquad (2.65)$$

and its borderline expansion is

$$\text{nPX:} \quad t_B \approx t_i - \sqrt{2a_i^3}\left(\frac{1}{3} - \frac{a_i K}{20} + \frac{3a_i^2 K^2}{224}\right). $$

Velocity profile given at time t_i

Given the velocity distribution $b_i(M)$ at time t_i, the hyperbolic case again has a direct solution for t_B, once (2.63) is applied,

$$\text{HX:} \quad t_B = t_i - K^{-3/2}\left[\sqrt{\left(\frac{b_i^2 + K}{b_i^2 - K}\right)^2 - 1} - \text{arcosh}\left(\frac{b_i^2 + K}{b_i^2 - K}\right)\right], \qquad (2.66)$$

for which the existence conditions and borderline expansion are

$$\text{HX:} \quad b_i > 0, \quad 0 < K < b_i^2, \quad \text{and} \quad t_i - t_B > \frac{4}{3}b_i^3, \qquad (2.67)$$

$$\text{nPX:} \quad t_B \approx t_i - \frac{4}{b_i^3}\left(\frac{1}{3} + \frac{2K}{5b_i^2} + \frac{3K^2}{7b_i^4}\right). $$

3. **Models with a simultaneous crunch time**

For elliptic models, the requirement of only decaying modes is that the crunch time must be simultaneous (i.e. R–W like),

$$t_C = \text{specified constant}, \qquad (2.68)$$

where (2.9) is the relation between t_C and t_B. H and P models with a simultaneous crunch time obviously have no bang, and are therefore collapsing at all times, and are the time-reverse of models with a simultaneous bang.

Density profile given at time t_i

The equations to be solved here are (2.45), (2.46) and (2.47) with (2.68), which give:

$$\text{EX:} \quad 0 = -2\pi + \arccos(1 - a_i x) - \sqrt{1 - (1 - a_i x)^2}$$
$$+ x^{3/2}(t_C - t_i) \stackrel{\text{def}}{=} \psi_{CDX}(x); \qquad (2.69)$$

$$\text{EC:} \quad 0 = -\arccos(1 - a_i x) + \sqrt{1 - (1 - a_i x)^2}$$
$$+ x^{3/2}(t_C - t_i) \stackrel{\text{def}}{=} \psi_{CDC}(x); \qquad (2.70)$$

$$\text{HC:} \quad 0 = \sqrt{(1 + a_i x)^2 - 1} - \text{arcosh}(1 + a_i x)$$
$$- x^{3/2}(t_C - t_i) \overset{\text{def}}{=} \psi_{CDH}(x); \tag{2.71}$$

and the conditions for each particular solution are:

$$\text{EX:} \quad \pi \left(\frac{a_i}{2}\right)^{3/2} < t_C - t_i, \tag{2.72a}$$

$$\text{EC:} \quad \sqrt{2}\,\frac{a_i^{3/2}}{3} < t_C - t_i < \pi \left(\frac{a_i}{2}\right)^{3/2}, \tag{2.72b}$$

$$\text{HC:} \quad t_C - t_i < \sqrt{2}\,\frac{a_i^{3/2}}{3}. \tag{2.72c}$$

The equations for the borderline cases are:

$$\text{nPC:} \quad x \approx \frac{20}{3a_i}\left(-\frac{\text{d}\tau_P}{\tau_P} + \frac{25\text{d}\tau_P^2}{14\tau_P^2}\right); $$

$$\text{nEM:} \quad x \approx \frac{2}{a_i} - \frac{\text{d}\tau_N^2}{a_i^4} + \frac{3\pi\text{d}\tau_N^3}{2^{5/2}a_i^{11/2}}; $$

where $\tau_i = t_C - t_i$, while (2.57) defines τ_P, τ_N and $\text{d}\tau_N$. Once the above are solved, E and t_B then follow from (2.19b) and (2.68).

Velocity profile given at time t_i

These models are found by solving

$$\text{EX:} \quad 0 = -2\pi + \arccos\left(\frac{b_i^2 - x}{b_i^2 + x}\right) - \sqrt{1 - \left(\frac{b_i^2 - x}{b_i^2 + x}\right)^2}$$
$$+ x^{3/2}(t_C - t_i) \overset{\text{def}}{=} \psi_{CVX}(x); \tag{2.73}$$

$$\text{EC:} \quad 0 = -\arccos\left(\frac{b_i^2 - x}{b_i^2 + x}\right) + \sqrt{1 - \left(\frac{b_i^2 - x}{b_i^2 + x}\right)^2}$$
$$+ x^{3/2}(t_C - t_i) \overset{\text{def}}{=} \psi_{CVC}(x); \tag{2.74}$$

$$\text{HC:} \quad 0 = \sqrt{\left(\frac{b_i^2 + x}{b_i^2 - x}\right)^2 - 1} - \text{arcosh}\left(\frac{b_i^2 + x}{b_i^2 - x}\right)$$
$$- x^{3/2}(t_C - t_i) \overset{\text{def}}{=} \psi_{CVH}(x), \tag{2.75}$$

and the following hold for each solution to exist:

$$\text{EX:} \quad b_i > 0 \quad \text{and} \quad t_C > t_i, \tag{2.76a}$$

$$\text{EC:} \quad b_i < 0 \quad \text{and} \quad t_C - t_i < \frac{4}{3}b_i^3, \tag{2.76b}$$

$$\text{HC:} \quad b_i < 0 \quad \text{and} \quad t_C - t_i > \frac{4}{3}b_i^3. \tag{2.76c}$$

The borderline series are given by:

$$\text{nPX:} \quad x \approx \frac{5b_i^2}{6}\left(\frac{\mathrm{d}\tau_P}{\tau_P} - \frac{25\mathrm{d}\tau_P^2}{28\tau_P^2}\right);$$

$$\text{nEM:} \quad x \approx x_N\left(1 + \frac{16}{3\pi^2}\frac{b_i}{\bar{b}} + \frac{64}{3\pi^4}\frac{b_i^2}{\bar{b}^2}\right);$$

where

$$\tau_i = t_C - t_i, \qquad (2.77)$$

while (2.62) define τ_P, x_N and $\bar{b}$.

4. **Growing and decaying modes**

For pure decaying modes, hyperbolic (and parabolic) models that are expanding must become R–W like at late times, whereas for elliptic models and for hyperbolic and parabolic models that are collapsing, they must become R–W like at the crunch.

Conversely, for pure growing modes, hyperbolic models that are collapsing must have been R–W like in the distant past, while elliptic models and expanding hyperbolic and parabolic models must have been R–W like at the bang.

This is summarised in the following table:

	only growing modes	only decaying modes
HX, PX	$t_{B,r} = 0$	$2E = KM^{2/3}$
E	$t_{B,r} = 0$	$t_{C,r} = 0$
PC, HC	$2E = KM^{2/3}$	$t_{C,r} = 0$

5. **Models with a simultaneous time of maximum expansion**

Not infrequently, authors seeking a manifestly regular initial condition for an inhomogeneous matter distribution have required a finite density distribution and zero initial velocity. This is achieved in an L–T model if the moment of maximum expansion occurs at the same time along all worldlines.

Naturally, only elliptic (EX and EC) models have a moment of maximum expansion, as follows from (2.8) and (2.9), and the condition that it be simultaneous is

$$t_{SMX} = t_B + \frac{\pi M}{(-2E)^{3/2}} = \text{specified constant.} \qquad (2.78)$$

This must be combined with another condition to obtain a solution.

Density profile given at time t_i

The equations to be solved, for the two relevant evolution types are

$$\text{EX:} \quad 0 = -\pi + \arccos(1 - a_i x) - \sqrt{1 - (1 - a_i x)^2}$$
$$+ x^{3/2}(t_{SMX} - t_i) \stackrel{\text{def}}{=} \psi_{SDX}(x) \tag{2.79}$$

$$\text{EC:} \quad 0 = \pi - \arccos(1 - a_i x) + \sqrt{1 - (1 - a_i x)^2}$$
$$+ x^{3/2}(t_{SMX} - t_i) \stackrel{\text{def}}{=} \psi_{SDC}(x), \tag{2.80}$$

and after numerical solution, E and t_B then follow from (2.19b) and (2.78). The above solutions exist when the following conditions hold:

$$\text{EX:} \quad t_{SMX} > t_i, \tag{2.81a}$$
$$\text{EC:} \quad t_{SMX} < t_i. \tag{2.81b}$$

Density profile given at a simultaneous time of maximum expansion

Although this is a special case of the above, it has an explicit solution. Given $\rho = \rho_{SMX}(M)$ at $t = t_{SMX}$, we calculate $R_{SMX}(M)$ using (2.18). From (2.5a) R has a maximum value $M/(-E)$, so we can write

$$(-E) = \frac{M}{R_{SMX}} \quad \Leftrightarrow \quad x = \frac{2}{a_{SMX}} \tag{2.82}$$

and use it in (2.78), giving the direct solution

$$t_B = t_{SMX} - \pi\sqrt{\frac{a_{SMX}^3}{8}} \, , \qquad E = -\frac{M^{2/3}}{a_{SMX}}, \tag{2.83}$$

which exists if

$$a_{SMX} > 0 \quad \text{and} \quad M > 0. \tag{2.84}$$

Velocity profile given at time t_i

The two elliptic cases are found from

$$\text{EX:} \quad 0 = -\pi + \arccos\left(\frac{b_i^2 - x}{b_i^2 + x}\right) - \sqrt{1 - \left(\frac{b_i^2 - x}{b_i^2 + x}\right)^2}$$
$$+ x^{3/2}(t_{SMX} - t_i) \stackrel{\text{def}}{=} \psi_{SVX}(x), \tag{2.85}$$

$$\text{EC:} \quad 0 = \pi - \arccos\left(\frac{b_i^2 - x}{b_i^2 + x}\right) + \sqrt{1 - \left(\frac{b_i^2 - x}{b_i^2 + x}\right)^2}$$
$$+ x^{3/2}(t_{SMX} - t_i) \stackrel{\text{def}}{=} \psi_{SVC}(x), \tag{2.86}$$

and they exist when

$$\text{EX:}\quad t_{SMX} > t_i, \tag{2.87a}$$

$$\text{EC:}\quad t_{SMX} < t_i. \tag{2.87b}$$

6. Models with given density and velocity profiles at the same time

For this case, we have both $a_i(M)$ and $b_i(M)$ at the same t_i, so we solve (2.2) for E,

$$2E = (R_{,t})_i^2 - \frac{2M}{R_i} \quad\Leftrightarrow\quad \pm x = \frac{2E}{M^{2/3}} = b_i^2 - \frac{2}{a_i}, \tag{2.88}$$

and obtain t_B from one of (2.44)–(2.47). The equation for t_B is sensitive to the sign of b_i (i.e., $R_{,t}$), even though the E equation is not. The conditions for existence of solutions are

$$\text{HX:}\quad b_i^2 > \frac{2}{a_i} \quad\text{and}\quad b_i > 0, \tag{2.89a}$$

$$\text{EX:}\quad b_i^2 < \frac{2}{a_i} \quad\text{and}\quad b_i > 0, \tag{2.89b}$$

$$\text{EC:}\quad b_i^2 < \frac{2}{a_i} \quad\text{and}\quad b_i < 0, \tag{2.89c}$$

$$\text{HC:}\quad b_i^2 > \frac{2}{a_i} \quad\text{and}\quad b_i < 0. \tag{2.89d}$$

An example of this method is given in Bolejko *et al.* (2005). The borderline cases need no special treatment, as there are no numerical difficulties arising from being close to them. The model is parabolic if $b_i^2 = 2/a_i$, and is at maximum expansion (at t_i) if $b_i = 0$.

7. Models with a given velocity profile at late times

By 'late times' we mean the asymptotic future, i.e. the limit $\eta \to \infty$ and $t \to \infty$, so this subsection applies only to expanding hyperbolic (HX) models. The time reverse – a collapsing hyperbolic (HC) model with a given density or velocity profile in the infinite past – follows in an obvious way.

From (2.7) and (2.2) we find

$$\lim_{\eta \to \infty} \frac{R}{t - t_B} = \sqrt{2E}\,, \qquad \lim_{t \to \infty} R_{,t} = \sqrt{2E}, \tag{2.90}$$

which gives simply

$$E = R_{,t}{}^2_{\text{late}}/2 \quad\Leftrightarrow\quad x = b_{\text{late}}^2. \tag{2.91}$$

This always exists, and fully determines $E(M)$, leaving $t_B(M)$ free to be determined by a second requirement.

Density profile given at time t_i

Since x is known from (2.91), we find t_B from

$$\text{HX:} \quad t_B = t_i - b_{\text{late}}^{-3}\left[\sqrt{(1 + a_i b_{\text{late}}^2)^2 - 1} - \text{arcosh}(1 + a_i b_{\text{late}}^2)\right], \quad (2.92)$$

which has a solution only if

$$\text{HX:} \quad b_{\text{late}} > 0 \quad \text{and} \quad a_i > 0. \quad (2.93)$$

The borderline case is given by

$$\text{nPX:} \quad t_B \approx t_i - \tau_P\left(1 - \frac{3 a_i b_{\text{late}}^2}{20} + \frac{9 a_i^2 b_{\text{late}}^4}{224}\right). $$

Velocity profile given at time t_i

The equation for t_B is

$$\text{HX:} \quad t_B = t_i - b_{\text{late}}^{-3}\left[\sqrt{\left(\frac{b_i^2 + b_{\text{late}}^2}{b_i^2 - b_{\text{late}}^2}\right)^2 - 1} - \text{arcosh}\left(\frac{b_i^2 + b_{\text{late}}^2}{b_i^2 - b_{\text{late}}^2}\right)\right], \quad (2.94)$$

which has a solution if only

$$\text{HX:} \quad b_i > b_{\text{late}} > 0. \quad (2.95)$$

The borderline case is given by

$$\text{nPX:} \quad t_B \approx t_i - \tau_P\left(1 + \frac{6 b_{\text{late}}^2}{5 b_i^2} + \frac{9 b_{\text{late}}^4}{7 b_i^4}\right). $$

Simultaneous bang time specified

Equations (2.52) and (2.91) fully specify the L–T arbitrary functions $E(M)$ and $t_B(M)$. The only existence condition is

$$b_{\text{late}} > 0. \quad (2.96)$$

8. **Models with a given density profile at late times**
   From (2.7) and (2.3) we obtain, using $\partial R/\partial M = R_{,r}/M_{,r}$,

$$\frac{\partial R/\partial M}{(t - t_B)} = \sqrt{2E}\left[\left(\frac{1}{M} - \frac{\text{d}E/\text{d}M}{E}\right)\frac{(\cosh\eta - 1)}{(\sinh\eta - \eta)}\right.$$

$$\left. + \left(\frac{3\,\text{d}E/\text{d}M}{2E} - \frac{1}{M}\right)\frac{\sinh\eta}{(\cosh\eta - 1)}\right]$$

$$- \frac{(2E)^2\,\text{d}t_B/\text{d}M}{M}\frac{\sinh\eta}{(\cosh\eta - 1)(\sinh\eta - \eta)}, \quad (2.97)$$

$$\kappa\rho(t - t_B)^3 = \frac{2}{[R/(t - t_B)]^2\,[(\partial R/\partial M)/(t - t_B)]}. \quad (2.98)$$

Assuming dt_B/dM is finite, we find that

$$\lim_{\eta \to \infty} \frac{\partial R/\partial M}{(t - t_B)} = \frac{dE/dM}{\sqrt{2E}}, \tag{2.99}$$

which with (2.90) leads to

$$\lim_{\eta \to \infty} \kappa\rho(t - t_B)^3 = \frac{2}{\sqrt{-2E}\, dE/dM}. \tag{2.100}$$

Clearly $\rho(t - t_B)^3$ freezes out – becomes time-independent – and

$$\lim_{\eta \to \infty} \kappa\rho(t - t_B)^3 = \frac{6}{(d/dM)\left((2E)^{3/2}\right)}. \tag{2.101}$$

Thus, if we specify the late time limit $[\rho(t - t_B)^3]_{\text{late}}(M)$, we have

$$x_{\text{late}}^{3/2} = \frac{(2E)^{3/2}}{M} = \frac{1}{M}\int_0^M \frac{6}{\kappa[\rho(t - t_B)^3]_{\text{late}}}\, d\widetilde{M}. \tag{2.102}$$

This assumes an origin exists, i.e. $E = 0$ at $M = 0$. If not, then we must specify some $E = E_i$ at some $M = M_i$.

Again, (2.91) fully determines $E(M)$, leaving $t_B(M)$ free.

Density profile given at time t_i

In this case we find t_B from

$$\text{HX:} \quad t_B = t_i - x_{\text{late}}^{-3/2}\left[\sqrt{(1 + a_i x_{\text{late}})^2 - 1} - \text{arcosh}(1 + a_i x_{\text{late}})\right], \tag{2.103}$$

with the existence condition

$$[\rho(t - t_B)^3]_{\text{late}} > 0, \tag{2.104}$$

and the borderline case is given by

$$\text{nPX:} \quad t_B \approx t_i - \tau_P\left(1 - \frac{3a_i x_{\text{late}}}{20} + \frac{9a_i^2 x_{\text{late}}^2}{224}\right). $$

Velocity profile given at time t_i

The equation for t_B is (x_1 stands for x_{late})

$$\text{HX:} \quad t_B = t_i - x_1^{-3/2}\left[\sqrt{\left(\frac{b_i^2 + x_1}{b_i^2 - x_1}\right)^2 - 1} - \text{arcosh}\left(\frac{b_i^2 + x_1}{b_i^2 - x_1}\right)\right], \tag{2.105}$$

with the existence condition

$$\text{HX:}\ b_i^2 > x_1, \tag{2.106}$$

and the borderline case is given by

$$\text{nPX:} \quad t_B \approx t_i - \tau_P \left(1 + \frac{6x_1}{5b_i^2} + \frac{9x_1^2}{7b_i^4} \right).$$

Simultaneous bang time specified

Again, (2.52) and (2.102) complete the specification of the L–T arbitrary functions and the solution always exists.

9. **Existence**

When constructing solutions, we need to know not only whether solutions exist, but also which type of evolution is applicable along each constant M worldline. We summarise the conditions for each case in Table 2.1, the derivations can be found in Hellaby and Krasiński (2006). The borderline conditions for the parabolic (P) and the maximum expansion (EM) cases are sufficiently obvious that they are not included. A few cases are omitted because they are time reverses of included cases, such as models that evolve from a given density profile in the infinite past to a simultaneous crunch.

We have developed several new ways of specifying the 'boundary' data needed to uniquely determine the evolution of an L–T model, and thus provided more options for designing models with particular properties or behaviours. Thus one can now easily generate models that start from an initial stationary state, or have only growing modes, or approach a specified density or velocity profile in the asymptotic future, or approach Friedmann models at late times or diverge from them, etc., as listed at the beginning of this section. The foregoing properties are combined in pairs to fully specify a particular L–T model. Although several of the individual properties considered here have previously been used, what is significant here is the equations that result from the many combinations of pairs of properties, and the derivation of existence conditions for the 3 types of solution. Also, our results have been presented in a form that is easily converted into coding for numerical calculations. Later work (Bolejko and Hellaby, 2008) provides an example of the use of some of the new L–T specification methods to create and evolve a model of the Shapley concentration and the Great Attractor.

2.1.7 Defining a model by initial and final profiles with $\Lambda > 0$

Initial density profile to a final density profile

In order to determine the evolution in the L–T model we need two initial conditions. In this subsection, these initial conditions will be the density profiles defined at two different instants: t_1 and t_2. In consequence of observational

Table 2.1. *Summary of L–T model specifications. The columns are: the two specifications to be used, the type of evolution, the equations to be solved for the arbitrary functions, and the conditions under which the pair of specifications has the listed evolution type.*

Spec 1	Spec 2	Ev	Eq no		Exist cond
Den prfl given	Den prfl given	HX:	(2.24),	(2.22)	(2.25)
'	'	EX:	(2.26),	(2.45)	(2.28)
'	'	EC:	(2.29),	(2.46)	(2.30)
Vel prfl given	Den prfl given	HX:	(2.31),	(2.22)	(2.32)
'	'	EX:	(2.33),	(2.45)	(2.34)
'	'	EC:	(2.36),	(2.46)	(2.35)
Vel prfl given	Vel prfl given	HX:	(2.37),	(2.48)	(2.38)
'	'	EX:	(2.40),	(2.49)	(2.41)
'	'	EC:	(2.43),	(2.50)	(2.42)
Sim bang	Den prfl given	HX:	(2.52),	(2.53)	(2.56a)
'	'	EX:	(2.52),	(2.54)	(2.56b)
'	'	EC:	(2.52),	(2.55)	(2.56c)
'	Vel prfl given	HX:	(2.52),	(2.58)	(2.61a)
'	'	EX:	(2.52),	(2.59)	(2.61b)
'	'	EC:	(2.52),	(2.60)	(2.61c)
Hom late	Den prfl given	HX:	(2.63),	(2.64)	(2.65)
'	Vel prfl given	HX:	(2.63),	(2.66)	(2.67)
Sim crunch	Den prfl given	EX:	(2.68),	(2.69)	(2.72a)
'	'	EC:	(2.68),	(2.70)	(2.72b)
'	'	HC:	(2.68),	(2.71)	(2.72c)
'	Vel prfl given	EX:	(2.68),	(2.73)	(2.76a)
'	'	EC:	(2.68),	(2.74)	(2.76b)
'	'	HC:	(2.68),	(2.75)	(2.76c)
Sim max ex	Den prfl given	EX:	(2.78),	(2.79)	(2.81a)
'	'	EC:	(2.78),	(2.80)	(2.81b)
'	Den prfl at SMX	EM:	(2.83)		(2.84)
'	Vel prfl given	EX:	(2.78),	(2.85)	(2.87a)
'	'	EC:	(2.78),	(2.86)	(2.87b)
Sim den and	Vel prfl given	HX:	(2.88),	(2.44)	(2.89a)
'	'	EX:	(2.88),	(2.45)	(2.89b)
'	'	EC:	(2.88),	(2.46)	(2.89c)
'	'	HC:	(2.88),	(2.47)	(2.89d)
Late vel	Den prfl given	HX:	(2.91),	(2.92)	(2.93)
'	Vel prfl given	HX:	(2.91),	(2.94)	(2.95)
'	Sim bang	HX:	(2.91),	(2.52)	(2.96)
Late den	Den prfl given	HX:	(2.102),	(2.103)	(2.104)
'	Vel prfl given	HX:	(2.102),	(2.105)	(2.106)
'	Sim bang	HX:	(2.102),	(2.52)	always

implications, we are interested in joining these profiles only in the expansion phase, i.e. $t_2 > t_1 \Rightarrow R_2 > R_1$, with a positive value of the cosmological constant, $\Lambda > 0$.

To determine the evolution from (2.2) we need to know E, M, R_i and Λ. The function $M(r)$ is chosen to be the radial coordinate:

$$r' = M(r). \tag{2.107}$$

From the given initial and final density distributions and from (2.3) we can derive the functions $R_1 = R(t_1, M)$ and $R_2 = (t_2, M)$:

$$R_{1,2}^3 = \int_0^M dM' \left. \frac{6}{\kappa\rho} \right|_{t=t_{1,2}} , \qquad (2.108)$$

where we assume that $R(M = 0) = 0$.

In order to calculate E, (2.4) can be re-written for the two instants t_1 and t_2. Then subtracting one equation from another:

$$\Phi(E) = \int_{R_1}^{R_2} \frac{du}{\sqrt{2E + 2M/u + \frac{1}{3}\Lambda u^2}} = t_2 - t_1. \qquad (2.109)$$

If there exists a solution, i.e. if $\phi(u) \overset{\text{def}}{=} 2E + 2M/u + \frac{1}{3}\Lambda u^2 \geq 0$ (for any $u \in [R_1, R_2]$), then:

$$\frac{\partial\Phi(E)}{\partial E} = -\int_{R_1}^{R_2} \frac{du}{\left(2E + 2M/u + \frac{1}{3}\Lambda u^2\right)^{3/2}} \leqslant 0. \qquad (2.110)$$

So $\Phi(E)$ is a monotonically decreasing function of E. Hence, if a solution $E((t_2 - t_1))$ of (2.109) exists, then it is unique.

To consider the existence of a solution, the range of values of Φ must be examined. Since

$$\lim_{E \to +\infty} \Phi(E) = 0,$$

this means that when $(R_2 - R_1)$ is large and the time interval is short, the solution will exist. When the value of R_2 is close to R_1 and the time interval is large, then there could be problems.

A solution of (2.109) will exist only if $\phi(u) \geq 0$ in the whole range of integration. The function $\phi(u)$ has one minimum at $u_c = (3M/\Lambda)^{1/3}$. This naturally divides the integration into three cases:

1. $R_1 \leq R_2 \leq u_c$:
   In the interval $u \in [R_1, u_c]$, ϕ is a monotonically decreasing function of u, so its minimum value in the range of integration is at $u = R_2$. Given R_2, E cannot be smaller than the limiting value E_l, at which $\phi(R_2)|_{E=E_l} = 0$, so

$$2E \geq 2E_l \overset{\text{def}}{=} -\frac{2M}{R_2} - \frac{1}{3}\Lambda R_2^2.$$

   Let us consider if there exists a solution when E approaches the limiting value E_l. Then the following inequality can be written:

$$2E_l + \frac{2M}{u} + \frac{1}{3}\Lambda u^2 \geq \left(-\frac{2M}{R_2^2} + \frac{2}{3}\Lambda R_2\right)u + \frac{2M}{R_2} - \frac{2}{3}\Lambda R_2^2 \geq 0. \qquad (2.111)$$

The first inequality follows from the fact that ϕ is a convex function,

$$\phi_{,uu} = \frac{4M}{u^3} + \frac{2}{3}\Lambda > 0,$$

and a tangent to $\phi(u) - \phi(R_2)$ at $u = R_2$ is:

$$f(u) = \left(-\frac{2M}{R_2{}^2} + \frac{2}{3}\Lambda R_2\right)u + \frac{2M}{R_2} - \frac{2}{3}\Lambda R_2{}^2. \qquad (2.112)$$

The second inequality follows from the construction of the tangent, which at the point $u = R_2$ is zero and, having negative slope, must be positive for $u < R_2$.

Using (2.111) we can prove that $\Phi(E)$ converges, even if $E \to E_l$:

$$\int_{R_1}^{R_2} \frac{\mathrm{d}u}{\sqrt{2E_l + \frac{2M}{u} + \frac{1}{3}\Lambda u^2}} \leq \int_{R_1}^{R_2} \frac{\mathrm{d}u}{\sqrt{\left(-\frac{2M}{R_2{}^2} + \frac{2}{3}\Lambda R_2\right)u + \frac{2M}{R_2} - \frac{2}{3}\Lambda R_2{}^2}}$$

$$= 2\sqrt{\left(-\frac{2M}{R_2{}^2} + \frac{2}{3}\Lambda R_2\right)u + \frac{2M}{R_2} - \frac{2}{3}\Lambda R_2{}^2}$$

$$\times \left(-\frac{2M}{R_2{}^2} + \frac{1}{3}\Lambda R_2\right)^{-1}\Bigg|_{u=R_1}^{u=R_2} < \infty. \qquad (2.113)$$

This means that there exists a limiting (maximum) interval Δt_l:

$$\lim_{E \to E_l} \Phi(E) = \Delta t_l = \int_{R_1}^{R_2} \frac{\mathrm{d}u}{\sqrt{-\frac{2M}{R_2} - \frac{1}{3}\Lambda R_2{}^2 + \frac{2M}{u} + \frac{1}{3}\Lambda u^2}}. \qquad (2.114)$$

Two density distributions (R_1 and R_2) can be joined by an L–T evolution only if $t_2 - t_1 \leq \Delta t_l$.

2. $u_c \leq R_1 \leq R_2$:

In the interval $u \in [u_c, R_2]$ the function $\phi(u)$ is monotonically increasing, so its minimum value in the range of integration is at $u = R_1$. Consequently, now the limiting value $E = E_r$ is determined by $\phi(R_1)|_{E=E_r} = 0$, i.e.

$$2E \geq 2E_r = -\frac{2M}{R_1} - \frac{1}{3}\Lambda R_1{}^2.$$

Using the same arguments as above it can be proved that:

$$\lim_{E \to E_r} \Phi(E) = \Delta t_r = \int_{R_1}^{R_2} \frac{\mathrm{d}u}{\sqrt{-\frac{2M}{R_1} - \frac{1}{3}\Lambda R_1{}^2 + \frac{2M}{u} + \frac{1}{3}\Lambda u^2}}. \qquad (2.115)$$

This implies that two density distributions (determining R_1 and R_2) can be joined by an L–T evolution only if $t_2 - t_1 \leq \Delta t_r$.

3. $R_1 \leq u_c \leq R_2$:

In the interval $u \in [R_1, R_2]$, ϕ has a minimum at u_c. Now $\phi(u) \geq 0$ implies that E must be greater than the limiting value, E_c:

$$2E_c = -\left(9\Lambda M^2\right)^{1/3}.$$

Since in both intervals, $u \leq u_c$ and $u \geq u_c$, Φ does not diverge, there exists a limiting (maximum) interval Δt_c:

$$\lim_{E \to E_c} \Phi(E) = \Delta t_c = \int_{R_1}^{R_2} \frac{du}{\sqrt{-\left(9\Lambda M^2\right)^{1/3} + 2M/u + \frac{1}{3}\Lambda u^2}}. \tag{2.116}$$

So, two density distributions (defined by R_1 and R_2) can be joined only if $t_2 - t_1 \leq \Delta t_c$.

The above analysis implies that, as in the case $\Lambda = 0$, not all density profiles defined at different instants can be joined with an expanding L–T evolution. There exists an upper limit for the time interval $t_2 - t_1$ beyond which these density profiles cannot be connected. This means that if, for a given R_1 and R_2, the time interval $(t_2 - t_1)$ is too large, then $R(t_2) > R_2$.

If the value of $t_2 - t_1$ is smaller than that limit interval, (2.109) has only one solution and it can be solved for any instants t_1 and t_2, hence E can be calculated. The above analysis is valid if $-1/2 < E$, but in models of structure formation, values close to $\mathrm{E} = -1/2$ are not expected to occur. Once E is found, t_B can be calculated from (2.4) and the L–T model is determined. By solving (2.2), $R(r,t)$ can be found, and from (2.3) the density distribution can be found for any instant.

Initial velocity profile to a final density profile

In this subsection we examine if there exists a solution of the evolution equations from a given initial velocity to a final density profile. Because of the intended application to the observed Universe, we are interested in joining these profiles only in the expansion phase, i.e. $t_2 > t_1 \Rightarrow R_2 > R_1$, with a positive value of the cosmological constant, $\Lambda > 0$.

As above, the evolution of the system is calculated from (2.2), therefore the values of E, R_i, M and Λ must be derived from the initial conditions. $M(r)$ is again chosen to be the radial coordinate. From the final density distribution and from (2.3) we can calculate R_2. E can be derived from the initial velocity distribution:

$$R_{,t}(t_1, M) = V(M),$$
$$2E(M) = V^2(M) - \frac{2M}{R_1} - \frac{1}{3}\Lambda R_1^{\,2}, \tag{2.117}$$

In order to derive R_1 let us write an equation analogous to (2.109):

$$\Psi(u) = \int_u^{R_2} \frac{dy}{\sqrt{V^2 - 2M/u - \frac{1}{3}\Lambda u^2 + 2M/y + \frac{1}{3}\Lambda y^2}} = t_2 - t_1, \tag{2.118}$$

where $u = R_1$; we use u instead of R_1 to emphasise that R_1 is now an unknown variable.

If there exists a solution, i.e. if $\psi(y) = V^2 - 2M/u - \frac{1}{3}\Lambda u^2 + 2M/y + \frac{1}{3}\Lambda y^2 \geq 0$, then we can write:

$$\frac{\partial \Psi(u)}{\partial u} = \int_u^{R_2} \frac{\left(\frac{1}{3}\Lambda u - M/u^2\right) dy}{\left(V^2 - 2M/u - \frac{1}{3}\Lambda u^2 + 2M/y + \frac{1}{3}\Lambda y^2\right)^{3/2}} - \frac{1}{V}. \qquad (2.119)$$

Since $\left(\frac{1}{3}\Lambda u - M/u^2\right)$ is a monotonically increasing function of u (if $\Lambda > 0$):

$$\frac{\partial \Psi(u)}{\partial u} \leqslant \int_u^{R_2} \frac{\left(\frac{1}{3}\Lambda y - M/y^2\right) dy}{\left(V^2 - 2M/u - \frac{1}{3}\Lambda u^2 + 2M/y + \frac{1}{3}\Lambda y^2\right)^{3/2}} - \frac{1}{V}$$

$$= - \left. \frac{1}{\sqrt{V^2 - 2M/u - \frac{1}{3}\Lambda u^2 + 2M/y + \frac{1}{3}\Lambda y^2}} \right|_{y=u}^{|y=R_2} - \frac{1}{V}$$

$$= - \frac{1}{R_{,t}(t_2, M)} < 0. \qquad (2.120)$$

So $\Psi(u)$ is a monotonically decreasing function, and if there exists a solution of (2.118), it is unique and determines R_1.

As in the previous case (density to density), ψ has one minimum at $y_c = (3M/\Lambda)^{1/3}$. Since R_1 is an unknown quantity, to examine the value of Ψ we have to consider only two cases:

1. $R_2 \leq y_c$.
   From the condition $\psi(y) \geq 0$ and from the fact that in the interval $y \in [R_1, y_c]$, ψ is a monotonically decreasing function of y, there exists a minimum value of $u = u_m$, which can be derived from

$$V^2 - 2M/u_m - \frac{1}{3}\Lambda u_m^{\,2} + 2M/R_2 + \frac{1}{3}\Lambda R_2^{\,2} = 0. \qquad (2.121)$$

Then, repeating the same argument as in the previous case, it can be proved that $\Psi(u)$ has an upper limit:

$$\sup \Psi(u) = \Psi(u_m) = \Delta t_m = \int_{u_m}^{R_2} \frac{dy}{\sqrt{V^2 - \frac{2M}{u_m} - \frac{1}{3}\Lambda u_m^{\,2} + \frac{2M}{y} + \frac{1}{3}\Lambda y^2}}. \qquad (2.122)$$

So, only if $t_2 - t_1 \leq \Delta t_m$, can (2.118) be solved and the value of R_1 calculated.

2. $R_2 \geq y_c$.
   In this interval, the minimum value of $R_1 = u_m$ can be derived from the condition $\psi(y) \geq 0$ at the point $y = y_c$ where ψ has the minimum:

$$V^2 - 2M/u_m - \frac{1}{3}\Lambda u_m^{\,2} + \left(9\Lambda M^2\right)^{1/3} = 0. \qquad (2.123)$$

Then it can be shown that

$$\sup \Psi(u) = \Psi(u_m) = \Delta t_m = \int\limits_{u_m}^{R_2} \frac{dy}{\sqrt{V^2 - \frac{2M}{u_m} - \frac{1}{3}\Lambda u_m{}^2 + \frac{2M}{y} + \frac{1}{3}\Lambda y^2}}.$$

(2.124)

As in the case of two density profiles, not all initial velocity profiles can be joined by an expanding L–T evolution to a given final density profile. There exists an upper limit for the time interval $t_2 - t_1$ beyond which these profiles cannot be connected. So, only if $t_2 - t_1 \leq \Delta t_m$, can (2.118) be solved and the value of R_1 calculated.

When R_1 is calculated, $E(M)$ can be found from (2.117). Once E is found, t_B can be calculated from (2.4) and the L–T model is determined. By solving (2.2), $R(r,t)$ is found, and from (2.3) the density distribution can be found for any instant.

2.1.8 A unified new form of the L–T solution

Tanimoto and Nambu (2007) have recently proposed a new representation of the L–T solution, which, in a formal sense, exhibits the same form whatever the sign of the $E(r)$ function. We give here an account of this work since its results will be used in Sec. 5.4.

First, they observe that, in the $E > 0$ case, (2.7b) shows that η can be regarded as a function of $2E[(t - t_B)/M]^{2/3}$. Therefore, from (2.7a), the function R can be written

$$R(t,r) = \frac{M}{2E}X\left[2E\left(\frac{t - t_B}{M}\right)^{2/3}\right],$$

(2.125)

where a function $X(x)$ has been introduced. To make the $E \to 0$ limit regular, x is then factored out from the $X(x)$ function and a function S is defined by $X(x) \equiv 6^{1/3}xS(-6^{-2/3}x)$, which gives

$$R(t,r) = [6M(r)]^{1/3}[t - t_B(r)]^{2/3}S\left[-2E(r)\left(\frac{t - t_B(r)}{6M(r)}\right)^{2/3}\right],$$

(2.126)

where the numerical factors are a convention.

The same observation is applicable to both the $E < 0$ and $E = 0$ cases, and, as a result, (2.126) provides a desirable form for $R(t,r)$, explicit in terms of t and r, for which S changes smoothly as E changes sign. In practice, though, S must be expressed parametrically.

To check the consistency of this solution, it must be verified that the function $S(x)$ is smooth at $x = 0$. First we note that S is a function of the parameter $\zeta = (-E/|E|)\eta^2$ [where η is the parameter from (2.5)–(2.7)] and that in practice its form is different for the different cases: $x < 0$ (corresponding to $E > 0$), $x > 0$

(corresponding to $E < 0$) and $x = E = 0$. However, $\mathcal{S}$ and x expansions in powers of ζ exhibit the same form for all three cases, which is

$$\mathcal{S}(x) = \left(\frac{3}{4}\right)^{1/3} \left(1 - \frac{\zeta}{20} + \frac{\zeta^2}{1680} + \mathcal{O}(\zeta^3)\right), \qquad (2.127)$$

$$x = 6^{-4/3} \left(\zeta - \frac{\zeta^2}{30} + \frac{\zeta^3}{25200} + \mathcal{O}(\zeta^4)\right).$$

Since both $\mathcal{S}$ and x are smooth (actually, analytic) functions of ζ, in particular in the neighbourhood of $\zeta = 0$, this shows that $\mathcal{S}(x)$ is smooth at $x = 0$ as desired.

In the following, we summarise some useful properties of the function $\mathcal{S}(x)$ such as they are presented in Tanimoto and Nambu (2007).

First, substituting (2.126) into (2.2) without the cosmological constant, one obtains the following first-order ordinary differential equation for $\mathcal{S}(x)$

$$\frac{4}{3}[\mathcal{S}(x) + x\mathcal{S}_{,x}(x)]^2 + 3x - \frac{1}{\mathcal{S}(x)} = 0, \qquad (2.128)$$

where $\mathcal{S}_{,x} \equiv d\mathcal{S}/dx$.

Now, $\mathcal{S}(x)$ is not the only function satisfying (2.128). This equation admits other solutions. To single out $\mathcal{S}(x)$ which is smooth at $x = 0$, it is shown that it is the unique solution of (2.128) that 'traverses' $x = 0$, i.e. $\mathcal{S}_{,x}(x)$ is finite at $x = 0$. Note on the other hand that the term $x\mathcal{S}_{,x}(x)$ in (2.128) vanishes at $x = 0$, provided $\mathcal{S}_{,x}(x)$ is finite there, and therefore, (2.128) degenerates into an algebraic equation at $x = 0$. This equation uniquely determines $\mathcal{S}(0)$ which is found to be $\mathcal{S}(0) = (3/4)^{1/3}$.

Remember that (2.2) without the cosmological constant is an integral of the second-order Einstein equation (Lemaître, 1933; Tolman, 1934)

$$2RR_{,tt} + R_{,t}^2 - 2E = 0, \qquad (2.129)$$

from which one obtains the equation for $\mathcal{S}(x)$,

$$x[2\mathcal{S}(x)\mathcal{S}_{,xx}(x) + \mathcal{S}_{,x}^2(x)] + 5\mathcal{S}(x)\mathcal{S}_{,x}(x) + \frac{9}{4} = 0. \qquad (2.130)$$

For $x \neq 0$, this equation is useful in case one wants to eliminate derivatives of $\mathcal{S}(x)$ of order higher than the first. Moreover, one can use this equation to determine $\mathcal{S}_{,x}(0)$ from $\mathcal{S}(0)$ and find $\mathcal{S}_{,x}(0) = -(9/20)\mathcal{S}(0)$. Similarly, taking higher derivatives of (2.130), one can successively compute the values of arbitrarily high derivatives of $\mathcal{S}(x)$ at $x = 0$ in terms of $\mathcal{S}(0)$. One of the useful applications of this property is to obtain a series expansion of $\mathcal{S}(x)$ about $x = 0$. With the value of $\mathcal{S}(0)$ determined above, the first terms of this expansion read

$$\mathcal{S}(x) = \left(\frac{3}{4}\right)^{1/3} \left[1 - \frac{3}{5}\left(\frac{3}{4}\right)^{1/3} x - \frac{27}{350}\left(\frac{9}{2}\right)^{1/3} x^2 + \mathcal{O}(x^3)\right]. \qquad (2.131)$$

An interesting one-parameter special solution of (2.130) is

$$\widetilde{\mathcal{V}}_\beta(x) = \sqrt{-x} - \frac{\beta}{x}, \tag{2.132}$$

where β is an arbitrary constant parameter. It can be shown that $\widetilde{\mathcal{V}}_\beta(x) \to \mathcal{S}(x)$ when $x \to -\infty$. In particular, $\widetilde{\mathcal{V}}_0(x) = \sqrt{-x}$ approaches $\mathcal{S}(x)$ from below and can be useful for various estimates of $\mathcal{S}(x)$ for large $-x$.

Finally, it is worth mentioning that the function

$$\mathcal{S}_C(x) \equiv \sqrt{x_C - x}, \tag{2.133}$$

which is not a solution of any of the above differential equations, provides a good approximation of $\mathcal{S}(x)$ for the whole domain $x \leq x_C$. The function $\mathcal{S}_C(x)$ asymptotically approaches $\mathcal{S}(x)$ from above as $x \to -\infty$.

A last elementary estimate establishes

$$x\mathcal{S}_{,x}(x) < \frac{\mathcal{S}(x)}{2}. \tag{2.134}$$

2.2 The Lemaître model

The evolution of a spherically symmetric perfect fluid was first considered by Lemaître (1933).[2] The metric in comoving coordinates is of the form:

$$ds^2 = e^{A(t,r)}dt^2 - e^{B(t,r)}dr^2 - R^2(t,r)\left(d\vartheta^2 + \sin^2\vartheta d\varphi^2\right). \tag{2.135}$$

The Einstein equations reduce to

$$\kappa R^2 R_{,r}\,\rho = 2M_{,r}, \tag{2.136}$$

$$\kappa R^2 R_{,t}\,p = -2M_{,t}, \tag{2.137}$$

where M is defined by

$$2M(t,r) = R(t,r) + R(t,r)e^{-A(t,r)}R_{,t}^2(t,r)$$
$$- e^{-B(t,r)}R_{,r}^2(t,r)R(t,r) - \frac{1}{3}\Lambda R^3(t,r). \tag{2.138}$$

In the Newtonian limit, Mc^2/G is equal to the mass inside the shell of radial coordinate r. However, it is not an integrated rest mass, but the active gravitational mass that generates the gravitational field. As can be seen from (2.138),

---

[2] In the literature, the name 'Lemaître model' is not used for the medium we consider here; indeed we propose now to introduce this name. The consideration described in this section is usually credited to Misner and Sharp (1964), and occasionally to Podurets (1964). However, it was Lemaître (1933) who first presented it, and the main credit should belong to him. In fact, Lemaître's consideration was still more general – he considered a fluid with anisotropic pressure.

the mass is not constant in time, and in the expanding Universe it decreases, as seen from (2.137).

From the equations of motion $T^{\alpha\beta}{}_{;\beta} = 0$ we obtain:

$$T^{0\alpha}{}_{;\alpha} = 0 \Rightarrow B_{,t} + 4\frac{R_{,t}}{R} = -\frac{2\rho_{,t}}{\rho + p}, \qquad (2.139)$$

$$T^{1\alpha}{}_{;\alpha} = 0 \Rightarrow A_{,r} = -\frac{2p_{,r}}{\rho + p}, \qquad (2.140)$$

$$T^{2\alpha}{}_{;\alpha} = T^{3\alpha}{}_{;\alpha} = 0 \Rightarrow \frac{\partial p}{\partial \theta} = 0, \quad \frac{\partial p}{\partial \phi} = 0. \qquad (2.141)$$

Equation (2.141) states the well-known fact that the perfect fluid energy–momentum tensor inherits the symmetries of the metric of the spacetime.

The function B can be written in the following form:

$$B(t,r) = B(t_0, r) + \ln R_{,r}{}^2 - \int_{t_0}^{t} d\tilde{t}\frac{A_{,r}}{R_{,r}}\frac{R_{,t}}{}. \qquad (2.142)$$

Using (2.140) we obtain:

$$e^{B(t,r)} = \frac{R_{,r}{}^2(t,r)}{1 + 2E(r)} \exp\left(\int_{t_0}^{t} d\tilde{t}\frac{2R_{,t}(\tilde{t}, r)}{[\rho(\tilde{t}, r) + p(\tilde{t}, r)] R_{,r}(\tilde{t}, r)}p_{,r}(\tilde{t}, r)\right), \qquad (2.143)$$

where $E(r)$ is an arbitrary function.

In the special case of dust, the above equations reproduce the Lemaître–Tolman model. As follows from (2.140) and (2.143), if $p_{,r} = 0$ then $e^A = 1$, $e^B = R_{,r}{}^2/(1 + 2E)$. Then the metric (2.135) becomes (2.1).

2.3 The Szekeres solution

2.3.1 Definition and metric

In the following, the Szekeres solution will be used to describe the formation and evolution of galaxy clusters and voids and to study light propagation effects acting on the CMB photons in a lumpy Universe. We begin by presenting the most basic properties of this solution – those that will be useful in the next chapters. A more extended account can be found in Plebański and Krasiński (2006) and Hellaby and Krasiński (2002, 2008).

The metric of the Szekeres solution is

$$ds^2 = dt^2 - e^{2\alpha}dr^2 - e^{2\beta}\left(dx^2 + dy^2\right), \qquad (2.144)$$

where α and β are functions of (t, x, y, r) to be determined from the Einstein equations with a dust source. The coordinates of (2.144) are comoving so that $u^\mu = \delta^\mu{}_0$.

Several different parametrisations are in use for the solutions of the Einstein equations with the metric (2.144), and some notations are conflicting. We shall present it in a parametrisation based on the original Szekeres (1975a, 1975b) papers, and in one later-introduced parametrisation that will be of use in some of our considerations.

There are in fact two families of Szekeres solutions, depending on whether $\beta_{,r} = 0$ or $\beta_{,r} \neq 0$. The first family is a simultaneous generalisation of the Friedmann and Kantowski–Sachs (1966) models. Since so far it has found no useful application in astrophysical cosmology, we shall not discuss it here (see Plebański and Krasiński, 2006). After the Einstein equations are solved, the metric functions in the second family become

$$e^{\beta} = \Phi(t, r)e^{\nu(r, x, y)},$$

$$e^{\alpha} = h(r)\Phi(t, r)\beta_{,r} \equiv h(r)\left(\Phi_{,r} + \Phi\nu_{,r}\right),$$

$$e^{-\nu} = A(r)\left(x^2 + y^2\right) + 2B_1(r)x + 2B_2(r)y + C(r), \qquad (2.145)$$

where the function $\Phi(t, r)$ is a solution of the equation

$$\Phi_{,t}{}^2 = -k(r) + \frac{2M(r)}{\Phi} + \frac{1}{3}\Lambda\Phi^2; \qquad (2.146)$$

while $h(r)$, $k(r)$, $M(r)$, $A(r)$, $B_1(r)$, $B_2(r)$ and $C(r)$ are arbitrary functions obeying

$$g(r) \overset{\text{def}}{=} 4\left(AC - B_1{}^2 - B_2{}^2\right) = 1/h^2(r) + k(r). \qquad (2.147)$$

The mass density in energy units is

$$\kappa\rho = \frac{\left(2Me^{3\nu}\right)_{,r}}{e^{2\beta}\left(e^{\beta}\right)_{,r}}; \qquad \kappa = 8\pi G/c^4. \qquad (2.148)$$

As in the L–T model, the bang time function follows from (2.146):

$$\int_0^{\Phi} \frac{\mathrm{d}\widetilde{\Phi}}{\sqrt{-k + 2M/\widetilde{\Phi} + \frac{1}{3}\Lambda\widetilde{\Phi}^2}} = t - t_B(r). \qquad (2.149)$$

The Szekeres metric has in general no symmetry, but acquires a 3-dimensional symmetry group with 2-dimensional orbits when A, B_1, B_2 and C are all constant (that is, when $\nu_{,r} = 0$).

The sign of $g(r)$ determines the geometry of the constant t, constant r 2-surfaces (and the symmetry of the constant A, B_1, B_2 and C case). The geometry of these surfaces is spherical, planar or hyperbolic (pseudo-spherical) when $g > 0$, $g = 0$ or $g < 0$, respectively. With A, B_1, B_2 and C being functions of r, the surfaces $r = $ const within a single space $t = $ const may have different geometries, i.e. they can be spheres in one part of the space and surfaces of constant negative curvature elsewhere, the curvature being zero at the boundary.

The sign of $k(r)$ determines the type of evolution: with $k > 0 = \Lambda$ the model expands away from an initial singularity and then recollapses to a final singularity; with $k < 0 = \Lambda$ the model is either ever-expanding or ever-collapsing, depending on the initial conditions; $k = 0$ is the intermediate case corresponding to the 'flat' Friedmann model ($k = 0$ can also occur on a 3-surface as the boundary between a region with $k > 0$ and another one with $k < 0$). The sign of $k(r)$ influences the sign of $g(r)$. Since $1/h^2$ in (2.147) must be non-negative,[3] we have the following: with $g > 0$ (spherical geometry), all three types of evolution are allowed; with $g = 0$ (plane geometry), k must be non-positive (only parabolic or hyperbolic evolutions are allowed); and with $g < 0$ (hyperbolic geometry), k must be strictly negative, so only the hyperbolic evolution is allowed.

The Friedmann limit follows when $\Phi(t, r) = rR(t)$, $k = k_0 r^2$ where $k_0 = \text{const}$ and $B_1 = B_2 = 0$, $C = 4A = 1$. This definition of the Friedmann limit includes the definition of the limiting radial coordinate [the Szekeres model is covariant with the transformations $r = f(r')$, where $f(r')$ is an arbitrary function].

The Szekeres models are subdivided according to the sign of $g(r)$ into the quasi-spherical ones (with $g > 0$), quasi-plane ($g = 0$) and quasi-hyperbolic ($g < 0$). Despite suggestions to the contrary made in the literature, the geometry of the latter two classes has not been investigated at all and is not really understood; see Hellaby and Krasiński (2008) and Krasiński (2008) for recent work on their interpretation. Only the quasi-spherical model has been rather well investigated, and found useful application in astrophysical cosmology, so in this text we limit ourselves to this class. The quasi-spherical model may be imagined as a generalisation of the L–T model in which the spheres of constant mass are made non-concentric. The functions $A(r)$, $B_1(r)$ and $B_2(r)$ determine how the centre of a sphere changes its position in a space $t = \text{const}$ when the radius of the sphere is increased or decreased. Still, this is a rather simple geometry because all the arbitrary functions depend on one variable, r.

Often, it is more practical to reparametrise the arbitrary functions in the Szekeres metric as follows (this parametrisation was invented by Hellaby, 1996b). Even if $A = 0$ initially, a transformation of the (x, y)-coordinates can restore $A \neq 0$, so we may assume $A \neq 0$ with no loss of generality (see Plebański and Krasiński, 2006). Then let $g \neq 0$. Writing $A = \sqrt{|g|}/(2S)$, $B_1 = -\sqrt{|g|}P/(2S)$, $B_2 = -\sqrt{|g|}Q/(2S)$, $\varepsilon \overset{\text{def}}{=} g/|g|$, $k = |g|\widetilde{k}$ and $\Phi = \sqrt{|g|}\widetilde{\Phi}$, we can represent the metric (2.145) as[4]

$$e^{-\nu} = \sqrt{|g|}\mathcal{E}, \qquad \mathcal{E} \overset{\text{def}}{=} \frac{S}{2}\left[\left(\frac{x - P}{S}\right)^2 + \left(\frac{y - Q}{S}\right)^2 + \varepsilon\right],$$

$$ds^2 = dt^2 - \frac{\left(\Phi_{,r} - \Phi\mathcal{E}_{,r}/\mathcal{E}\right)^2}{\varepsilon - k(r)}dr^2 - \frac{\Phi^2}{\mathcal{E}^2}\left(dx^2 + dy^2\right), \qquad (2.150)$$

---

[3] $1/h(r)$ can be zero at isolated points – it is then either a coordinate singularity or a neck or belly – but not on open intervals.

[4] The tildes were dropped in (2.150) and in all further text for better readability. The Φ in (2.150) is in fact $\widetilde{\Phi}$ and the $k(r)$ is $\widetilde{k}(r)$. The redefinitions listed imply, via (2.147), $C = \sqrt{|g|}\left[\left(P^2 + Q^2\right)/S + \varepsilon S\right]/2$ and $h^2 = 1/[|g|(\varepsilon - \widetilde{k})]$. Also, the function M of (2.146) and (2.148) needs to be redefined by $M = \sqrt{|g|}^3 \widetilde{M}$. The M used from now on is in fact $\widetilde{M}$.

where $\varepsilon = +1$ for the quasi-spherical model. When $g = 0$, the transition from (2.145) to (2.150) is $A = 1/(2S)$, $B_1 = -P/(2S)$, $B_2 = -Q/(2S)$ and Φ is unchanged.[5] Then (2.150) applies with $\varepsilon = 0$, and the resulting model is quasi-plane, which we do not discuss here.

The parametrisation introduced above makes several formulae simpler, mainly because all the functions present in it are independent – the constraint (2.147) is identically fulfilled in it. However, this parametrisation obscures the fact, evident in (2.144)–(2.147), that *the same* Szekeres model may be quasi-spherical in one part of the spacetime, and quasi-hyperbolic elsewhere, with the boundary between these two regions being quasi-plane; see an explicit simple example in Hellaby and Krasiński (2008). In most of the literature published so far, these models have been considered separately, but this was either for purposes of systematic research, or with a specific application in view that fixed the sign of $g(r)$.

Within each single $\{t = \text{const}, r = \text{const}\}$ surface, the transformation from the (ϑ, φ) coordinates of (2.1) to the (x, y) coordinates is

$$(x - P, y - Q)/S = \cot(\vartheta/2)(\cos\varphi, \sin\varphi). \tag{2.151}$$

This transformation is called a **stereographic projection**. Its geometric interpretation is shown in Fig. 2.1.

We have briefly presented the Szekeres metric following the method by which it was originally derived: the form (2.144) was postulated, and then the complete solution of the Einstein equations was found. This definition is not covariant. Definitions based on invariant properties do exist, but are somewhat complicated (see Plebański and Krasiński, 2006).

2.3.2 General properties of the Szekeres solutions

The Szekeres models are solutions of the Einstein equations for an irrotational perfect fluid source. In addition the acceleration vanishes, $u^\alpha_{;\beta} u^\beta = 0$. The shear tensor is

$$\sigma^\alpha{}_\beta = \frac{1}{3}\left(\frac{\Phi_{,tr} - \Phi_{,t}\,\Phi_{,r}\,/\Phi}{\Phi_{,r} - \Phi\mathcal{E}_{,r}\,/\mathcal{E}}\right)\text{diag}(0, 2, -1, -1). \tag{2.152}$$

The scalar of expansion is

$$\theta = u^\alpha{}_{;\alpha} = \frac{\Phi_{,tr} + 2\Phi_{,t}\,\Phi_{,r}\,/\Phi - 3\Phi_{,t}\,\mathcal{E}_{,r}\,/\mathcal{E}}{\Phi_{,r} - \Phi\mathcal{E}_{,r}\,/\mathcal{E}}. \tag{2.153}$$

In the Friedmann limit, $R \to rR_F$, where R_F is the Friedmann scale factor. Thus $\theta \to 3H_0$, where H_0 is the current value of the Hubble parameter and $\sigma^\alpha{}_\beta \to 0$.

---

[5] The implied changes in C and h are then $C = (P^2 + Q^2)/(2S)$, $h^2 = -1/k$; k and M remain unchanged.

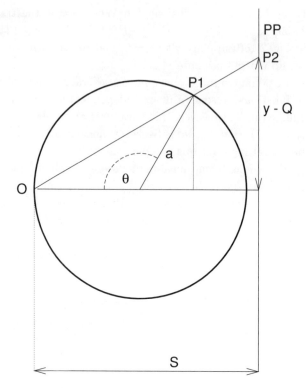

Fig. 2.1. The stereographic projection from the (ϑ, φ) coordinates to the Szekeres coordinates (x, y) for a sphere. The radius of the sphere is $a = 1$. A point P1 in the surface with the (ϑ, φ) coordinates is mapped to the point P2 in the plane PP with the (x, y) coordinates related to (ϑ, φ) by (2.151). The point O is the pole of projection; it lies on the straight line P1–P2. S is the distance from the pole to the plane. The figure is the cross section of the setup with the $x = P$ plane. The coordinate φ is the azimuthal angle around the axis of symmetry that passes through O.

The equations of motion $T^{\alpha\beta}{}_{;\beta} = 0$ reduce to the continuity equation:

$$\rho_{,t} + \rho\theta = 0. \tag{2.154}$$

In the expanding Universe, $\theta > 0$, so the density decreases. The structures which exist in the Universe emerged either due to slower expansion of the space (formation of overdense regions) or due to faster expansion (formation of underdense regions).

The Weyl curvature decomposed into its electric and magnetic parts is of the following form:

$$E^{\alpha}{}_{\beta} = C^{\alpha}{}_{\gamma\beta\delta}u^{\gamma}u^{\delta} = \frac{M(3\Phi_{,r} - \Phi M_{,r}/M)}{3\Phi^{3}(\Phi_{,r} - \Phi\mathcal{E}_{,r}/\mathcal{E})}\operatorname{diag}(0, 2, -1, -1),$$

$$H_{\alpha\beta} = \frac{1}{2}\sqrt{-g}\epsilon_{\alpha\gamma\mu\nu}C^{\mu\nu}{}_{\beta\delta}u^{\gamma}u^{\delta} = 0. \tag{2.155}$$

Finally the 4D and 3D Ricci scalars are

$$^4\mathcal{R} = -4\Lambda - 8\pi G\rho,$$

$$^3\mathcal{R} = 2\frac{k}{\Phi^2}\left(\frac{\Phi k_{,r}/k - 2\Phi\mathcal{E}_{,r}/\mathcal{E}}{\Phi_{,r} - \Phi\mathcal{E}_{,r}/\mathcal{E}} + 1\right). \tag{2.156}$$

2.3.3 Properties of the quasi-spherical Szekeres solution

When $\Lambda \neq 0$, the solutions of (2.146) involve elliptic functions. A general formal integral of (2.146) was found by Barrow and Stein-Schabes (1984). Any solution of (2.146) will contain one more arbitrary function of r, denoted $t_B(r)$, and will enter the solution in the combination $(t - t_B(r))$. The instant $t = t_B(r)$ defines the initial moment of evolution; when $\Lambda = 0$ it is necessarily a singularity corresponding to $\Phi = 0$, and it goes over into the Big Bang singularity in the Friedmann limit; in fact it is called Big Bang in the general case, too. When $t_{B,r} \neq 0$ (that is, in general) the instant of singularity is position-dependent, just as it is in the L–T model.

As for the L–T model, another singularity may occur where $(e^{\beta})_{,r} = 0$ (if this equation has solutions). This is a shell crossing, but now it is qualitatively different from that in the L–T model. If this singularity is present in the L–T model, then whole spherical shells collide there. Here, as can be seen from (2.145), the equation $(e^{\beta})_{,r} = 0$ may define a subset of the $\{x, y\}$ surface at each r. When a shell crossing exists, its intersection with a $t = \text{const}$ space will be a circle, or, in exceptional cases, a single point, in each of a set of constant r 2-surfaces, and not a sphere (see subsection 'Shell crossings' further on).

Equation (2.146), is formally identical to the Friedmann equation, but with k and M depending on r, so each surface $r = \text{const}$ evolves independently of the others. The solutions $\Phi(t, r)$ are the same as the L–T solutions $R(t, r)$ of (2.2), and are unaffected by the dependence of the Szekeres solutions on the (x, y) coordinates.

As defined by (2.145) and (2.146), the Szekeres models contain 8 functions of r, of which only 7 are arbitrary because they must obey (2.147). The parametrisation of (2.150) turns the function $g(r)$ to a constant parameter ε, thus reducing the number to 6. By a choice of r (still arbitrary up to now), we can fix one more function [for example, by defining $r' = M(r)$]. Thus, the number of arbitrary functions that correspond to physical degrees of freedom is 5.

The quasi-spherical Szekeres spacetime can be matched to the Schwarschild solution across an $r = \text{const}$ hypersurface (Bonnor, 1976).

In the following, we will represent the quasi-spherical Szekeres solution with $\beta_{,r} \neq 0$ in the parametrisation introduced in (2.150). The formula for density in these variables is

$$\kappa\rho = \frac{2\left(M_{,r} - 3M\mathcal{E}_{,r}/\mathcal{E}\right)}{\Phi^2\left(\Phi_{,r} - \Phi\mathcal{E}_{,r}/\mathcal{E}\right)}. \tag{2.157}$$

Basic physical restrictions

We choose $\Phi \geq 0$ ($\Phi = 0$ is an origin, the bang or the crunch; in no case is a continuation to negative Φ possible) and $M(r) \geq 0$, so that any vacuum exterior has positive Schwarzschild mass.

Since $(\mathrm{d}x^2 + \mathrm{d}y^2)/\mathcal{E}^2$ maps to the unit sphere, $|S(r)| \neq 0$ is needed for a sensible mapping, so $S > 0$ is a reasonable choice. For a well-behaved r-coordinate, we require

$$\infty > \frac{(\Phi_{,r} - \Phi \mathcal{E}_{,r} /\mathcal{E})^2}{1 - k} > 0, \tag{2.158}$$

i.e. $1 - k > 0$ except where $(\Phi_{,r} - \Phi \mathcal{E}_{,r} /\mathcal{E})^2 = 0$.

The density must be positive, and the Kretschmann scalar $R_{\alpha\beta\gamma\delta} R^{\alpha\beta\gamma\delta}$ must be finite, which adds

$$\text{either} \quad M_{,r} - 3M\mathcal{E}_{,r} /\mathcal{E} \geq 0 \quad \text{and} \quad \Phi_{,r} - \Phi\mathcal{E}_{,r} /\mathcal{E} \geq 0, \tag{2.159}$$

$$\text{or} \quad M_{,r} - 3M\mathcal{E}_{,r} /\mathcal{E} \leq 0 \quad \text{and} \quad \Phi_{,r} - \Phi\mathcal{E}_{,r} /\mathcal{E} \leq 0. \tag{2.160}$$

If $(\Phi_{,r} - \Phi\mathcal{E}_{,r} /\mathcal{E})$ passes through 0 anywhere other than at a regular extremum, we have a shell crossing.

The significance of $\mathcal{E}$

As seen from (2.150), with $\varepsilon = +1$ and $S > 0$, we always have $\mathcal{E} > 0$. Can $\mathcal{E}_{,r}$ change sign?

We will consider the variation of $\mathcal{E}(r, x, y)$ around the spheres of constant t and r. Setting $\varepsilon = +1$ and applying the transformation (2.151) to $\mathcal{E}$ in (2.150) and to its derivative we see that the locus of $\mathcal{E}_{,r} = 0$ is

$$S_{,r} \cos\vartheta + P_{,r} \sin\vartheta \cos\varphi + Q_{,r} \sin\vartheta \sin\varphi = 0. \tag{2.161}$$

Writing

$$Z = \cos\vartheta, \qquad Y = \sin\vartheta \cos\varphi, \qquad X = \sin\vartheta \sin\varphi, \tag{2.162}$$

we see that (X, Y, Z) is on a unit sphere through $(0, 0, 0)$, and (2.161) becomes $S_{,r} Z + P_{,r} X + Q_{,r} Y = 0$ which is the equation of an arbitrary plane through $(0, 0, 0)$. Such planes all intersect the unit sphere along great circles, therefore $\mathcal{E}_{,r} = 0$ is a great circle, with locus

$$\tan\vartheta = -S_{,r} / (P_{,r} \cos\varphi + Q_{,r} \sin\varphi). \tag{2.163}$$

With $S_{,r} > 0$, the function $\mathcal{E}_{,r}$ is negative on one side of the circle, zero on the circle and positive on the other side.

As seen from (2.163), with $S_{,r} = 0$ we have $\vartheta = 0$, which means that the great circle defined by $\mathcal{E}_{,r} = 0$ passes through the pole of stereographic projection. In this case, the image of the circle $\mathcal{E}_{,r} = 0$ on the (x, y) plane is a straight line passing through $(x, y) = (P, Q)$. $\mathcal{E}_{,r}$ has a different sign on each side of the straight line.

From the definition of $\mathcal{E}$ we find

$$\mathcal{E}_{,r}\,/\mathcal{E} = -\left[S_{,r}\cos\vartheta + \sin\vartheta\,(P_{,r}\cos\varphi + Q_{,r}\sin\varphi)\right]/S; \qquad (2.164)$$

thus

$$\mathcal{E}_{,r}\,/\mathcal{E} = \text{constant} \Rightarrow S_{,r}\,Z + P_{,r}\,X + Q_{,r}\,Y = S \times \text{constant}, \qquad (2.165)$$

which is a plane parallel to the $\mathcal{E}_{,r} = 0$ plane, implying that all loci of $\mathcal{E}_{,r}\,/\mathcal{E} = \text{constant}$ are circles parallel to the $\mathcal{E}_{,r} = 0$ great circle.

The extrema of $\mathcal{E}_{,r}\,/\mathcal{E}$ are located at

$$\tan\varphi_e = \frac{Q_{,r}}{P_{,r}}, \qquad (2.166)$$

$$\tan\vartheta_e = \frac{P_{,r}\cos\varphi_e + Q_{,r}\sin\varphi_e}{S_{,r}} = \varepsilon_1\frac{\sqrt{P_{,r}{}^2 + Q_{,r}{}^2}}{S_{,r}}, \qquad (2.167)$$

$$\cos\vartheta_e = \varepsilon_2\frac{S_{,r}}{\sqrt{S_{,r}{}^2 + P_{,r}{}^2 + Q_{,r}{}^2}}, \qquad \varepsilon_1, \varepsilon_2 = \pm 1. \qquad (2.168)$$

The extreme value is then

$$(\mathcal{E}_{,r}\,/\mathcal{E})_{\text{extreme}} = -\varepsilon_2\sqrt{S_{,r}{}^2 + P_{,r}{}^2 + Q_{,r}{}^2}/S. \qquad (2.169)$$

Since

$$(\sin\vartheta_e\cos\varphi_e, \sin\vartheta_e\sin\varphi_e, \cos\vartheta_e) = \frac{\varepsilon_2\,(P_{,r}\,, Q_{,r}\,, S_{,r}\,)}{\sqrt{P_{,r}{}^2 + Q_{,r}{}^2 + S_{,r}{}^2}},$$

the extreme values of $\mathcal{E}_{,r}\,/\mathcal{E}$ are poles to the great circles of $\mathcal{E}_{,r} = 0$.

Clearly $\mathcal{E}_{,r}\,/\mathcal{E}$ has a dipole variation around each constant-r sphere, changing sign when we go over to the antipodal point: $(\vartheta, \varphi) \to (\pi - \vartheta, \varphi + \pi)$. Writing

$$\Phi_{,r} - \Phi\mathcal{E}_{,r}\,/\mathcal{E} = \Phi_{,r} + \Phi\left[S_{,r}\cos\vartheta + \sin\vartheta\,(P_{,r}\cos\varphi + Q_{,r}\sin\varphi)\right]/S, \qquad (2.170)$$

we see that $\Phi\mathcal{E}_{,r}\,/\mathcal{E}$ is the correction to the radial separation $\Phi_{,r}$ of constant-r shells, due to their not being concentric. In particular, $\Phi S_{,r}\,/S$ is the forward $(\vartheta = 0)$ displacement, and $\Phi P_{,r}\,/S$ and $\Phi Q_{,r}\,/S$ are the two sideways displacements $(\vartheta = \pi/2, \varphi = 0)$ and $(\vartheta = \pi/2, \varphi = \pi/2)$. The shortest 'radial' distance is where $\mathcal{E}_{,r}\,/\mathcal{E}$ is maximum.

It will be shown further on that, where $\Phi_{,r} > 0$, $\mathcal{E}_{,r}\,/\mathcal{E} \leq M_{,r}\,/(3M)$ and $\mathcal{E}_{,r}\,/\mathcal{E} \leq \Phi_{,r}\,/\Phi$ are required to avoid shell crossings, and also $\Phi_{,r}\,/\Phi > M_{,r}\,/(3M)$ in (2.184). These inequalities, together with $M_{,r} > 0$, imply that the density given by (2.157), as a function of $x \stackrel{\text{def}}{=} \mathcal{E}_{,r}\,/\mathcal{E}$,

$$\rho = \frac{2M_{,r}}{\Phi^2\Phi_{,r}}\frac{1 - 3Mx/M_{,r}}{1 - \Phi x/\Phi_{,r}}, \qquad (2.171)$$

has a negative derivative by x:

$$\rho_{,x} = \frac{\Phi/\Phi_{,r} - 3M/M_{,r}}{(1 - \Phi x/\Phi_{,r})^2} \cdot \frac{2M_{,r}}{\Phi^2 \Phi_{,r}} < 0, \qquad (2.172)$$

so the density is minimum where $\mathcal{E}_{,r}/\mathcal{E}$ is maximum.

Conditions of regularity at the origin

We list here only those conditions of regularity that must be directly taken into account in considering the galaxy cluster and void models. The full list of conditions, with derivations, can be found in Hellaby and Krasiński (2002) and Plebański and Krasiński (2006).

At the origin of spherical coordinates (which we assume to be at $r = 0$; this can always be satisfied by a transformation of r), we have $\Phi(t, 0) = 0$, $\forall t$ (the 2-spheres have no size). Similarly, $\Phi_{,t}(t, 0) = 0 = \Phi_{,tt}(t, 0)$, etc. $\forall t$. There will be a second origin, at $r = r_o$ say, in any closed, regular, $k > 0$ model. The regularity conditions say that

$$\lim_{r \to 0} M = 0, \qquad \lim_{r \to 0} k = 0, \qquad (2.173)$$

$$0 < \lim_{r \to 0} \left(|k|^{3/2}/M \right) < \infty, \qquad \lim_{r \to 0} [3Mk_{,r}/(2M_{,r}k)] = 1. \qquad (2.174)$$

All of the conditions taken together imply that, near an origin,

$$M \sim \Phi^3, \quad k \sim -\Phi^2, \quad S \sim \Phi^n, \quad P \sim \Phi^n, \quad Q \sim \Phi^n, \quad n \geq 0. \quad (2.175)$$

The condition $\mathcal{E}_{,r}/\mathcal{E} \leq M_{,r}/(3M)$ that will be obtained in the next subsection implies that $n \leq 1$ near an origin.

Shell crossings

A shell crossing, if it exists, is the locus of zeros of the function $\chi \overset{\text{def}}{=} \mathcal{E}\Phi_{,r}/\Phi - \mathcal{E}_{,r}$. Now $\Phi_{,r} > 0$ and $\chi < 0$ cannot hold for all x and y. This would lead to $\mathcal{E}_{,r} > \mathcal{E}\Phi_{,r}/\Phi > 0$, and we know that $\mathcal{E}_{,r}$ cannot be positive at all x and y (see Hellaby and Krasiński, 2002). Hence, with $\Phi_{,r} > 0$, there must be a region in which $\chi > 0$. By a similar argument, $\Phi_{,r} < 0$ and $\chi > 0$ cannot hold for all x and y, so with $\Phi_{,r} < 0$ there must be a region in which $\chi < 0$.

Assuming $\Phi_{,r} > 0$, can χ be positive for all x and y? A calculation (Hellaby and Krasiński, 2002) shows that χ will have the same sign for all x and y (i.e. there will be no shell crossings) if and only if

$$\frac{\Phi_{,r}^2}{\Phi^2} > \frac{P_{,r}^2 + Q_{,r}^2 + S_{,r}^2}{S^2} \overset{\text{def}}{=} \Psi^2(r). \qquad (2.176)$$

If $\Phi_{,r}^2/\Phi^2 = \Psi^2$, then $\chi = 0$ at one x value, say x_{SS}. In this case, the shell crossing is a single point in the constant-(t, r) surface, i.e. a curve in a space of constant t and a 2-surface in spacetime.

If $\Phi_{,r}{}^2/\Phi^2 < \Psi^2$, then the locus of $\chi = 0$ is in general a circle (a straight line in the special case $S_{,r}/S = -\Phi_{,r}/\Phi$) in the (x,y) plane. The straight line is a projection onto the (x,y) plane of a circle on the sphere of constant t and r, and hence is not really a special case.

When $\Phi_{,r}{}^2/\Phi^2 < \Psi^2$, the set $\chi = 0$ is a circle with the centre at

$$(x_{\rm SC}, y_{\rm SC}) = \left(P - \frac{P_{,r}}{S_{,r}/S + \Phi_{,r}/\Phi}, Q - \frac{Q_{,r}}{S_{,r}/S + \Phi_{,r}/\Phi} \right), \qquad (2.177)$$

and with the radius $L_{\rm SC} = \sqrt{\delta}/(S_{,r}/S + \Phi_{,r}/\Phi)$, where

$$\delta \stackrel{\rm def}{=} P_{,r}{}^2 + Q_{,r}{}^2 + \varepsilon \left(S_{,r}{}^2 - S^2 \Phi_{,r}{}^2/\Phi^2 \right). \qquad (2.178)$$

This is in general a different circle from the one defined by $\mathcal{E}_{,r} = 0$. The shell-crossing set intersects with the surface of constant t and r along the line $\mathcal{E}_{,r}/\mathcal{E} = \Phi_{,r}/\Phi = $ const. As noted after (2.165), this is a circle that lies in a plane parallel to the $\mathcal{E}_{,r} = 0$ great circle. It follows that the $\mathcal{E}_{,r} = 0$ and SC circles cannot intersect unless they coincide.

Now we will consider the conditions for avoiding shell crossings. These were worked out by Szekeres (1975b) and improved upon by Hellaby and Krasiński (2002). The account here is based on the latter reference.

For positive density, (2.157) shows that $(M_{,r} - 3M\mathcal{E}_{,r}/\mathcal{E})$ and χ must have the same sign. Consider the case when both are positive. When $(M_{,r} - 3M\mathcal{E}_{,r}/\mathcal{E}) \leq 0$ and $\chi < 0$, the inequalities in all the following should be reversed.

Both $(M_{,r} - 3M\mathcal{E}_{,r}/\mathcal{E})$ and χ can be zero for a particular (x,y) value if $M_{,r}/3M = \Phi_{,r}/\Phi$, but the latter cannot hold for all time. This case can hold for all (x,y) only if $M_{,r} = 0$, $\mathcal{E}_{,r} = 0$ and $\Phi_{,r} = 0$, which requires all of $M_{,r}$, $k_{,r}$, $t_{B,r}$, $S_{,r}$, $P_{,r}$ and $Q_{,r}$ to be zero at some r value.

Consider the inequality $M_{,r} - 3M\mathcal{E}_{,r}/\mathcal{E} \geq 0$. It must hold for all values of $\mathcal{E}_{,r}/\mathcal{E}$, including the extreme value (2.169), for which

$$\frac{M_{,r}}{3M} \geq \left. \frac{\mathcal{E}_{,r}}{\mathcal{E}} \right|_{\rm max} = \frac{\sqrt{(S_{,r})^2 + (P_{,r})^2 + (Q_{,r})^2}}{S} \qquad \forall\, r. \qquad (2.179)$$

It is obvious that this is sufficient, and also that $M_{,r} \geq 0$ for all r.

We will now consider $\chi > 0$ for all three types of evolution. We quote the results only. For details see Hellaby and Krasiński (2002).

Hyperbolic evolution, $k < 0$

Here, $\chi > 0$ implies

$$t_{B,r} < 0 \qquad \forall\, r, \qquad (2.180)$$

$$\Phi_{,r}/\Phi - \mathcal{E}_{,r}/\mathcal{E} > 0 \qquad \Rightarrow \qquad k_{,r}/(2k) - \mathcal{E}_{,r}/\mathcal{E} > 0, \qquad (2.181)$$

$$k_{,r}/(2k) > \sqrt{(S_{,r})^2 + (P_{,r})^2 + (Q_{,r})^2}/S \qquad \forall\, r, \qquad (2.182)$$

which obviously implies $k_{,r} < 0 \ \forall \ r$. Since we already have $M_{,r} \geq 0$, this is sufficient, and implies $\Phi_{,r} > 0$.

Parabolic evolution, $k = 0$

All terms involving $k_{,r}/k$ cancel out and we retain conditions (2.180) and (2.179). Naturally, (2.182) ceases to impose any limit.

Elliptic evolution, $k > 0$

Here, the conditions are (2.180) and

$$\frac{2\pi M}{k^{3/2}} \left(\frac{M_{,r}}{M} - \frac{3k_{,r}}{2k} \right) + t_{B,r} > 0 \quad \forall \ r, \tag{2.183}$$

which says that the crunch time must increase with r. Since we already have $M_{,r} \geq 0$, these conditions are sufficient to keep $\Phi_{,r} > 0$ for all time. Further, it can be shown that

$$\Phi_{,r}/\Phi > M_{,r}/(3M), \tag{2.184}$$

so that (2.179) guarantees that for each given r the maximum of $\mathcal{E}_{,r}/\mathcal{E}$ as (x,y) are varied is no more than the minimum of $\Phi_{,r}/\Phi$ as η varies.

Although (2.183) implies $k_{,r}/(2k) < M_{,r}/(3M)$, a condition such as (2.182) is not needed in this case.

The mass-dipole

In the Szekeres solution discussed here, the distribution of mass over each single sphere $\{t = \text{const}, r = \text{const}\}$ has the form of a mass-dipole superposed on a monopole.[6] This was first noted by Szekeres (1975b) and then explained in much more detail by de Souza (1985). The presentation here is based on the latter reference, but somewhat modified (see also Plebański and Krasiński, 2006).

The basic idea is to separate the expression for matter density, (2.157), into a spherically symmetric part ρ_s, depending only on t and r, and a nonsymmetric $\Delta\rho$. Without additional requirements, this can be done in an infinite number of ways. The way to achieve a unique result is as follows. The calculation is simpler in the variables of (2.144)–(2.148), so we do it first in that representation. For better readability, we denote

$$\mathcal{N} = e^{-\nu}. \tag{2.185}$$

Adding and subtracting $H(t,r)/\Phi^2$ on the right-hand side of (2.148), where $H(t,r)$ is an arbitrary function, results in

$$\rho = \rho_s(t,r) + \Delta\rho(t,r,x,y), \tag{2.186}$$

---

[6] As will be seen from the following, it is possible to reduce this combination to a 'bare dipole' – but then the mass density would be negative in a part of the volume of the medium, and no-one knows how to interpret such an object physically.

where

$$\kappa \rho_s = \frac{H}{\Phi^2}, \qquad \kappa \Delta \rho = \frac{\mathcal{N}\,(2M_{,r} - H\Phi_{,r}) - \mathcal{N}_{,r}\,(6M - H\Phi)}{\Phi^2\,(\mathcal{N}\Phi_{,r} - \Phi\mathcal{N}_{,r})}. \qquad (2.187)$$

Now additional requirements on H will make the splitting unique. We transform (x, y) to spherical polar coordinates on a sphere of radius 1 that is tangent to the (x, y) plane at the point $(x, y) = (0, 0)$. The transformation is as in (2.151), but with $S = 1$ and $P = Q = 0$, viz.

$$x = \cot(\vartheta/2)\cos\varphi, \qquad y = \cot(\vartheta/2)\sin\varphi, \qquad (2.188)$$

which transforms $\mathcal{N} = e^{-\nu}$, as given by (2.145), to

$$\mathcal{N} = A\cot^2(\vartheta/2) + 2B_1\cot(\vartheta/2)\cos\varphi + 2B_2\cot(\vartheta/2)\sin\varphi + C. \qquad (2.189)$$

We substitute this in (2.187) and consider the equation $\Delta\rho = 0$. We embed the sphere in a Euclidean 3-space, and express (ϑ, φ) through the Cartesian coordinates in the space via (2.162). Using (2.189) and (2.162) in $\Delta\rho = 0$ we obtain

$$[A_{,r}\,(1 + Z) + 2B_{1,r}X + 2B_{2,r}Y + C_{,r}\,(1 - Z)]\,(6M - H\Phi)$$
$$= [A(1 + Z) + 2B_1 X + 2B_2 Y + C(1 - Z)]\,(2M_{,r} - H\Phi_{,r}). \qquad (2.190)$$

Now we require that the surface of $\Delta\rho = 0$ passes through the centre of the sphere, i.e. that (2.190) is fulfilled at $X = Y = Z = 0$, thus

$$(A + C)\,(2M_{,r} - H\Phi_{,r}) = (A_{,r} + C_{,r})\,(6M - H\Phi). \qquad (2.191)$$

Since (A, C, M, Φ) depend only on t and r, this can be solved for H:

$$H = \frac{2M_{,r}\,(A + C) - 6M(A + C)_{,r}}{\Phi_{,r}\,(A + C) - \Phi(A + C)_{,r}}. \qquad (2.192)$$

This solution makes sense except when $[(A + C)/\Phi]_{,r} \equiv 0$. But then, the Szekeres model would degenerate into the Friedmann model. Hence, (2.192) applies whenever the Szekeres model is inhomogeneous.

With H given by (2.192), (2.187) become

$$\kappa \rho_s = \frac{2M_{,r}\,(A + C) - 6M(A + C)_{,r}}{\Phi^2\,[\Phi_{,r}\,(A + C) - \Phi(A + C)_{,r}]}, \qquad (2.193)$$

$$\kappa \Delta \rho = \frac{A_{,r} + C_{,r} - (A + C)\mathcal{N}_{,r}/\mathcal{N}}{\Phi^2\,(\Phi_{,r} - \Phi\mathcal{N}_{,r}/\mathcal{N})}\,\frac{6M\Phi_{,r} - 2M_{,r}\Phi}{\Phi^2\,[\Phi_{,r}\,(A + C) - \Phi\,(A_{,r} + C_{,r})]}. \qquad (2.194)$$

Now, $\Delta\rho = 0$ has two solutions:

$$A_{,r} + C_{,r} - (A + C)\mathcal{N}_{,r}/\mathcal{N} = 0 \qquad (2.195)$$

and $(2M/\Phi^3)_{,r} = 0$. The second one defines a hypersurface that depends on t, that is, it is not comoving except when $(2M/\Phi^3)_{,r} \equiv 0$, in which case the matter density becomes spatially homogeneous. The first hypersurface, call it H_1, has its equation independent of t – it is a world-sheet of a comoving surface.

Moreover, $\Delta\rho$ changes sign when H_1 is crossed, and, in the variables (X, Y, Z), $\Delta\rho$ is antisymmetric with respect to H_1. Hence, $\Delta\rho$ is a dipole-like contribution to matter density. Although the separation (2.186) is global, the orientation of the dipole axis is different on every sphere $\{t = \text{const}, r = \text{const}\}$.[7]

Repeating the calculation in the variables of (2.150) we would obtain

$$\kappa\Delta\rho = \frac{\chi_{,r} - \chi\mathcal{E}_{,r}/\mathcal{E}}{\Phi^2\left(\Phi_{,r} - \Phi\mathcal{E}_{,r}/\mathcal{E}\right)} \frac{6M\Phi_{,r} - 2M_{,r}\,\Phi}{\Phi^2\left(\Phi_{,r}\,\chi - \Phi\chi_{,r}\right)}, \tag{2.196}$$

where

$$\chi \stackrel{\text{def}}{=} \frac{1 + P^2 + Q^2}{2S} + \frac{S}{2}. \tag{2.197}$$

The set H_1 intersects every $(t = \text{const}, r = \text{const})$ sphere along a circle, unless $P_{,r} = Q_{,r} = S_{,r} = 0 \,(= A_{,r} = C_{,r})$, in which case the dipole component of density is simply zero. The intersection of H_1 with any sphere of constant r is a circle parallel to the great circle $\mathcal{E}_{,r} = 0$, as noted after (2.165). It will coincide with the $\mathcal{E}_{,r} = 0$ circle at those points where $\chi_{,r} = 0$ (if they exist). The dipole-like component will be antisymmetric with respect to $\mathcal{E}_{,r}/\mathcal{E}$ only at those values of r where $\chi\Phi = 0 = \chi_{,r}\,\Phi_{,r}$, but such values may exist only at the centre, $\Phi = 0$, because (2.197) clearly implies $\chi > 0$.

---

[7] The 'bare dipole' results when $\rho_s = 0$, i.e. when $(A + C)/M^{1/3} = \text{const}$.

3
Light propagation

3.1 Light propagation in Lemaître–Tolman models

In Chapter 5, we will examine how a number of cosmological problems can be solved in the framework of the L–T cosmological models. Since the available observations are made on our past light cone, these analyses imply the use of differential equations determining the null geodesics and generally also the red-shift.

We therefore give below these sets of differential equations in both cases: of a central and of an off-centre observer. For the case of the on-centre observer, we give the equations as derived by Célérier (2000a) (see also Plebański and Krasiński, 2006, Sec. 18.11) and often used in the literature. For the case of the off-centre observer, two methods are described: the first can be found in Schneider and Célérier (1999) and the second in Alnes and Amarzguioui (2006).

3.1.1 Central observer

In this section, we describe the Universe by the $t > t_B(r)$ part of the (r, t) plane, increasing t corresponding to going from the past to the future.

In the geometric optics approximation, light travels on a light cone from a source to the observer. Therefore, by (2.1), a light ray issued from a radiating source with coordinates $(t, r, \vartheta, \varphi)$ and radially directed toward an observer located at the symmetry centre of the model satisfies

$$\frac{dt}{dr} = -\frac{R_{,r}(t, r)}{\sqrt{1 + 2E(r)}}. \tag{3.1}$$

We will denote the solution of this equation by $t_n(r)$.

Following the standard procedure (Célérier, 2000a; Plebański and Krasiński, 2006), we can calculate the redshift along a radial ray to be

$$\ln(1 + z) = \int_0^r \frac{R_{,tr'}(t, r')}{\sqrt{1 + 2E}} \, dr', \tag{3.2}$$

and then we can choose z as a parameter along the rays and obtain

$$\frac{dr}{dz} = \frac{\sqrt{1 + 2E(r)}}{(1 + z)R_{,tr}[t_n(r), r]}. \tag{3.3}$$

Equation (3.1) becomes therefore

$$\frac{\mathrm{d}t}{\mathrm{d}z} = -\frac{R_{,r}\left[t_n(r), r\right]}{(1+z)R_{,tr}\left[t_n(r), r\right]}. \tag{3.4}$$

Each null geodesic is a solution of (3.3)–(3.4), starting with z at the source and finishing at the observer with $z = 0$.

3.1.2 Off-centre observer. SC method

For an observer located off-centre, a possible singularity at the centre does not alter the validity of such a model. The calculations are more involved in this case, but they can be done numerically, and we give below one method allowing us to deal with this issue.

This method and the corresponding equations were established by Schneider and Célérier (1999) to study the dipole and quadrupole moments of the CMB in their 'Delayed Big-Bang' model (see Sec. 7.2) and were used by Célérier (2000b) to solve the horizon problem within a class of models with an off-centre observer.

An observer located at a distance r_p from the centre sees an axially symmetric universe with the symmetry axis passing through her and the centre because of the spherical symmetry of the model. It is thus legitimate to simplify the problem by integrating the geodesics in the meridional plane. The photon path is uniquely defined by the observer's position (t_p, r_p) and the angle α between the direction from which the light ray comes and the direction to the universe centre.

For the metric given by (2.1), the meridional plane is defined as $\vartheta = \pi/2$ ($\Longrightarrow$ $k^\vartheta = 0$), k^ϑ being the ϑ component of the photon wave-vector and $k^\mu \equiv \mathrm{d}x^\mu/\mathrm{d}\lambda$, where λ is an affine parameter along the null geodesics. For $\mu = t, r, \varphi$, we obtain, respectively,

$$\frac{\mathrm{d}t}{\mathrm{d}\lambda} = k^t, \tag{3.5}$$

$$\frac{\mathrm{d}r}{\mathrm{d}\lambda} = k^r = g^{rr}k_r = -\frac{k_r}{R_{,r}{}^2}, \tag{3.6}$$

$$\frac{\mathrm{d}\varphi}{\mathrm{d}\lambda} = k^\varphi = g^{\varphi\varphi}k_\varphi = -\frac{k_\varphi}{R^2}. \tag{3.7}$$

The geodesic equations of light,

$$\frac{\mathrm{d}^2 x^\mu}{\mathrm{d}\lambda^2} + \Gamma^\mu{}_{\nu\rho}\frac{\mathrm{d}x^\nu}{\mathrm{d}\lambda}\frac{\mathrm{d}x^\rho}{\mathrm{d}\lambda} = 0, \tag{3.8}$$

allow us to obtain, after some calculations:

$$\frac{\mathrm{d}k^t}{\mathrm{d}\lambda} = -\frac{R_{,tr}}{R_{,r}{}^3}(k_r)^2 - \frac{R_{,t}}{R^3}(k_\varphi)^2, \tag{3.9}$$

$$\frac{\mathrm{d}k_r}{\mathrm{d}\lambda} = -\frac{R_{,rr}}{R_{,r}{}^3}(k_r)^2 - \frac{R_{,r}}{R^3}(k_\varphi)^2, \tag{3.10}$$

$$k_\varphi = \text{const.} \tag{3.11}$$

For photons, $ds^2 = 0$ coupled with (3.5)–(3.7) gives

$$(k^t)^2 = \left(\frac{k_r}{R_{,r}}\right)^2 + \left(\frac{k_\varphi}{R}\right)^2 . \tag{3.12}$$

The equation for the redshift z of the source in comoving coordinates is (Ellis, 1971; Plebański and Krasiński, 2006)

$$1 + z = k^t / (k^t)_p , \tag{3.13}$$

k^t and $(k^t)_p$ being the time component of the photon wave-vector at the source and at the observer, respectively.

The set of five differential equations (3.5) to (3.10) can be integrated with the following initial conditions given at the observer:

$$t = t_p , \qquad r = r_p , \qquad (k^t)_p = 1.$$

With these conditions, the redshift of the source reads

$$1 + z = k^t . \tag{3.14}$$

We denote

$$R_p \equiv R(t_p, r_p), \qquad R'_p \equiv R_{,r}(t_p, r_p), \tag{3.15}$$

and so on. The observer at (t_p, r_p) sees the photon trajectory making an angle α with the direction toward the centre of the universe. Therefore we can write:

$$(k_r)_p = a \cos\alpha, \qquad (k_\varphi)_p = b \sin\alpha. \tag{3.16}$$

Substituting the former values of the components of $(k_\mu)_p$ into (3.12) written at (t_p, r_p), we find:

$$a = R'_p , \qquad b = R_p , \tag{3.17}$$

and (3.11) becomes

$$k_\varphi = R_p \sin\alpha. \tag{3.18}$$

Inserting this expression for k_φ into (3.12), we obtain an expression for k_r which can be substituted into the set of differential equations and yields, after some calculations, the reduced system of three differential equations for three unknown, t, r and k^t, which represents the null geodesics defining the past light cone of the observer:

$$\frac{dt}{d\lambda} = k^t , \tag{3.19}$$

$$\frac{dr}{d\lambda} = \pm \frac{1}{R_{,r}} \left[(k^t)^2 - \left(\frac{R_p \sin\alpha}{R}\right)^2 \right]^{1/2} , \tag{3.20}$$

$$\frac{dk^t}{d\lambda} = -\frac{R_{,tr}}{R_{,r}} (k^t)^2 + \left(\frac{R_{,tr}}{R_{,r}} - \frac{R_{,t}}{R}\right) \left(\frac{R_p \sin\alpha}{R}\right)^2 , \tag{3.21}$$

with a plus sign in (3.20) for an observer looking inward and a minus sign for an observer looking outward, provided the affine parameter λ is chosen to increase from $\lambda = 0$ at (t_p, r_p) to λ at the source (t, r). The redshift follows from (3.14).

3.1.3 Off-centre observer. AA method

This method was described in Alnes and Amarzguioui (2006) and used by them in Alnes and Amarzguioui (2007) to compute the diameter distance and the CMB anisotropies of the models studied in these articles. Although it is more complicated than the method proposed above in Sec. 3.1.2, we give an account of it for completeness.

The light cones are specified by (t, t_0, γ, ξ), where t_0 is the time when the photons hit the observer at polar angle γ and azimuthal angle ξ. For simplicity, and without any loss of generality, the z-axis is chosen to run from the origin through the observer. The spatial coordinates of the observer in the metric reference frame are $r = r_0$ and $\vartheta = 0$, φ being degenerate. A transformation of the coordinates for the photon trajectories from the observer frame back to the metric frame gives

$$t = t, \qquad r = \hat{r}(t, t_0, \gamma), \qquad \vartheta = \hat{\vartheta}(t, t_0, \gamma), \qquad \varphi = \xi, \qquad (3.22)$$

where $\hat{r}$ and $\hat{\vartheta}$ are solutions to the geodesic equations as functions of t and of the initial conditions r_0, t_0 and γ.

Owing to the axial symmetry about the z-axis, these functions do not depend on the azimuthal angle $\varphi = \xi$. Therefore, the photon trajectories are determined by only three of the geodesic equations (3.8), with $\mu = t, r, \vartheta$. For $\mu = t$, the equation reads

$$\frac{d^2 t}{d\lambda^2} + \frac{R_{,r} R_{,tr}}{1 + 2E} \left(\frac{dr}{d\lambda} \right)^2 + RR_{,t} \left(\frac{d\vartheta}{d\lambda} \right)^2 = 0. \qquad (3.23)$$

For $\mu = r$, the equation is

$$\frac{d^2 r}{d\lambda^2} + \left(\frac{R_{,rr}}{R_{,r}} - \frac{E_{,r}}{1 + 2E} \right) \left(\frac{dr}{d\lambda} \right)^2 + \frac{2R_{,tr}}{R_{,r}} \frac{dr}{d\lambda} \frac{dt}{d\lambda} - \frac{R(1 + 2E)}{R_{,r}} \left(\frac{d\vartheta}{d\lambda} \right)^2 = 0. \qquad (3.24)$$

Finally, for $\mu = \vartheta$,

$$\frac{d^2 \vartheta}{d\lambda^2} + \frac{2R_{,t}}{R} \frac{d\vartheta}{d\lambda} \frac{dt}{d\lambda} + \frac{2R_{,r}}{R} \frac{d\vartheta}{d\lambda} \frac{dr}{d\lambda} = 0, \qquad (3.25)$$

which can be written as a conservation equation for the angular momentum $J \equiv R^2 d\vartheta/d\lambda$,

$$\frac{d}{d\lambda} J = 0. \qquad (3.26)$$

In addition, the property of the photon 4-velocity, $k^\mu k_\mu = 0$, gives

$$-\left(\frac{dt}{d\lambda}\right)^2 + \frac{R_{,r}{}^2}{1+2E}\left(\frac{dr}{d\lambda}\right)^2 + R^2\left(\frac{d\vartheta}{d\lambda}\right)^2 = 0. \tag{3.27}$$

It is simplest to specify the initial conditions when the photons arrive at the observer as $t = t_0$, $r = r_0$ and $\vartheta = 0$. The photon trajectories hit the observer at an angle γ relative to the z-axis. The spatial components of the unit vector in the r, ϑ and φ directions along this z-axis are

$$v^i = \frac{\sqrt{1+2E}}{R_{,r}}(1,0,0). \tag{3.28}$$

Defining $u \equiv dt/d\lambda$ and $p \equiv dr/d\lambda$, we can write the components of the unit spatial vector in the direction of the photon path at $t = t_0$ as

$$u^i = \left|\frac{d\lambda}{dt}\right|\left(\frac{dr}{d\lambda}, \frac{d\vartheta}{d\lambda}, \frac{d\varphi}{d\lambda}\right) = -\frac{1}{u}(p, J/R^2, 0). \tag{3.29}$$

The minus sign in front of the r.h.s. of this equation shows that λ has been chosen to decrease with time, implying that the equations will be integrated backward on the past light cone of the observer.

The angle γ is given by the inner product of v^i and u^i such that

$$\cos\gamma = g_{ij}u^i v^j = -\frac{R_{,r}}{\sqrt{1+2E}}\frac{p}{u}. \tag{3.30}$$

Since the parametrisation of the photon path in terms of the affine parameter λ is determined up to linear transformations of λ, we can choose $\lambda = 0$ when $t = t_0$ and $u_0 \equiv u(\lambda = 0) = -1$. Using (3.27) and (3.30), these conditions become initial conditions for p and J:

$$p_0 = \frac{\sqrt{1+2E}}{R_{,r}}\cos\gamma, \qquad J_0 = J = R\sin\gamma. \tag{3.31}$$

Inserting u, p and J into the geodesic equations (3.23) and (3.24), and considering (3.26) for J, we obtain a system of three coupled first-order differential equations for the three unknown functions, with initial conditions given by $u_0 = -1$ and (3.31). They will be integrated under the constraint of (3.27), written as

$$-u^2 + \frac{R_{,r}{}^2}{1+2E}p^2 + \frac{J^2}{R^2} = 0. \tag{3.32}$$

Then the expressions for $t(\lambda)$, $r(\lambda)$ and $\vartheta(\lambda)$ follow straightforwardly. Note that we have somewhat simplified the method used by Alnes and Amarzguioui (2006 and 2007), by reducing the number of unknown functions and differential equations to be numerically integrated.

Now, to determine the redshift of the incoming photons, travelling on such geodesics, as a function of the direction γ, we consider a source emitting two photons with a time separation τ. The equation for the time coordinate along the first geodesic can be written $t_1(\lambda) = t(\lambda)$. The time along the second geodesic is therefore given by $t_2(\lambda) = t(\lambda) + \tau(\lambda)$. Both photons must satisfy (3.27). For the first photon, it reads

$$\left(\frac{\mathrm{d}t}{\mathrm{d}\lambda}\right)^2 = \frac{R_{,r}(t,r)^2}{1+2E}\left(\frac{\mathrm{d}r}{\mathrm{d}\lambda}\right)^2 + R(t,r)^2\left(\frac{\mathrm{d}\vartheta}{\mathrm{d}\lambda}\right)^2 = 0, \qquad (3.33)$$

while, for the second photon, this becomes

$$\left(\frac{\mathrm{d}(t+\tau)}{\mathrm{d}\lambda}\right)^2 = \frac{R_{,r}(t+\tau,r)^2}{1+2E}\left(\frac{\mathrm{d}r}{\mathrm{d}\lambda}\right)^2 + R(t+\tau,r)^2\left(\frac{\mathrm{d}\vartheta}{\mathrm{d}\lambda}\right)^2 = 0. \qquad (3.34)$$

Expanding (3.34) to first order in τ and using (3.33), we find

$$\frac{\mathrm{d}t}{\mathrm{d}\lambda}\frac{\mathrm{d}\tau}{\mathrm{d}\lambda} = \tau(\lambda)\left[\frac{R_{,r}\,R_{,tr}}{1+2E}\left(\frac{\mathrm{d}r}{\mathrm{d}\lambda}\right)^2 + RR_{,t}\left(\frac{\mathrm{d}\vartheta}{\mathrm{d}\lambda}\right)^2\right]. \qquad (3.35)$$

The redshift measured by the observer is defined as

$$1 + z(\lambda_e) \equiv \frac{\tau(\lambda_r)}{\tau(\lambda_e)}, \qquad (3.36)$$

where the subscripts r and e refer to the receiver and emitter positions, respectively. Differentiating (3.36) with respect to λ_e, we obtain

$$\frac{\mathrm{d}z}{\mathrm{d}\lambda_e} = -\frac{1}{\tau(\lambda_e)}\frac{\mathrm{d}\tau(\lambda_e)}{\mathrm{d}\lambda_e}\frac{\tau(\lambda_r)}{\tau(\lambda_e)}. \qquad (3.37)$$

Next, using (3.35) and (3.36), and suppressing the subscript e, we arrive at the equation

$$\frac{\mathrm{d}z}{\mathrm{d}\lambda} = -(1+z)\frac{\mathrm{d}\lambda}{\mathrm{d}t}\left[\frac{R_{,r}\,R_{,tr}}{1+2E}\left(\frac{\mathrm{d}r}{\mathrm{d}\lambda}\right)^2 + RR_{,t}\left(\frac{\mathrm{d}\vartheta}{\mathrm{d}\lambda}\right)^2\right]. \qquad (3.38)$$

Inserting the functions u, p and J into (3.38), we find a first-order differential equation for the redshift of a photon received by the observer today as a function of λ, which reads

$$\frac{\mathrm{d}\ln(1+z)}{\mathrm{d}\lambda} = -\frac{1}{u}\left[\frac{R_{,r}\,R_{,tr}}{1+2E}p^2 + \frac{R_{,t}}{R^3}J^2\right], \qquad (3.39)$$

with the initial condition $z(\lambda = 0) \equiv z_0 = 0$.

3.2 Apparent and event horizons in the L–T model

In modelling a black hole, horizons must be taken into account. Apparent and event horizons of L–T models were studied in Hellaby (1987), in which L–T models that generalise the Schwarzschild–Kruskal–Szekeres topology to non-vacuum were demonstrated. We lay out further details of the apparent horizon here, along the lines of Krasiński and Hellaby (2004b).

Let us write the evolution equation (2.2) with $\Lambda = 0$ as

$$R_{,t} = \ell\sqrt{\frac{2M}{R} + 2E}, \qquad \text{where} \quad \begin{cases} \ell = +1 & \text{in the expanding phase,} \\ \ell = -1 & \text{in the collapsing phase.} \end{cases} \qquad (3.40)$$

The radial light rays must be geodesics by symmetry. Their equation, found from (2.1), may be written as

$$\left.\frac{\mathrm{d}t}{\mathrm{d}r}\right|_n = \frac{jR_{,r}}{\sqrt{1 + 2E}}, \qquad \text{where} \quad \begin{cases} j = +1 & \text{for outgoing rays,} \\ j = -1 & \text{for incoming rays,} \end{cases} \qquad (3.41)$$

whose solution we write as $t = t_n(r)$ or just t_n. Similarly we write $R_n = R(t_n, r)$.

3.2.1 Apparent horizons

Along a ray, we have for the total derivative of the areal radius,

$$\frac{\mathrm{d}R_n}{\mathrm{d}r} = R_{,t}\left.\frac{\mathrm{d}t}{\mathrm{d}r}\right|_n + R_{,r} = \left(\ell j\frac{\sqrt{\frac{2M}{R} + 2E}}{\sqrt{1 + 2E}} + 1\right) R_{,r}. \qquad (3.42)$$

The apparent horizon (AH) is the hypersurface in spacetime on which the rays are momentarily at constant R:

$$\frac{\mathrm{d}R_n}{\mathrm{d}r} = 0 \qquad \Rightarrow \qquad \ell j = -1 \quad \text{and} \quad R = 2M. \qquad (3.43)$$

There are in fact two apparent horizons: the future AH (AH^+), where

$$j = +1 \quad \text{(outgoing rays),} \qquad \ell = -1 \quad \text{(in a collapsing phase),} \qquad (3.44)$$

and the past AH (AH^-), where

$$j = -1 \quad \text{(incoming rays),} \qquad \ell = +1 \quad \text{(in an expanding phase).} \qquad (3.45)$$

We find $\mathrm{d}t/\mathrm{d}r$ along the AH by differentiating (3.43):

$$R_{,t}\,\mathrm{d}t + R_{,r}\,\mathrm{d}r = 2M_{,r}\,\mathrm{d}r,$$

giving

$$\left.\frac{\mathrm{d}t}{\mathrm{d}r}\right|_{AH} = \frac{2M_{,r} - R_{,r}}{R_{,t}} = \frac{2M_{,r} - R_{,r}}{\ell\sqrt{2M/R + 2E}} = \frac{\ell(2M_{,r} - R_{,r})}{\sqrt{1 + 2E}}, \qquad (3.46)$$

since $R = 2M$ on the AH. In the vacuum case $\rho = 0$, which implies $M_{,r} = 0$, we have $\mathrm{d}t/\mathrm{d}r|_{AH} = \mathrm{d}t/\mathrm{d}r|_n$ since $\ell j = -1$. Note that $M_{,r} = 0$ could be only local, so the AH would only be null in that region. In the Schwarzschild metric, where $M_{,r} = 0$ everywhere, this is consistent with $R = 2M$ being the locus of the event horizons; and in this case they coincide with the apparent horizons.

Recall that in the Schwarzschild spacetime the future and past event horizons, EH^+ and EH^-, cross in the neck at the moment it is widest. (Call this set O.) This holds for AHs in L–T models with necks too. For hyperbolic regions, with $E \geq 0$ along each dust worldline, there is either only expansion or only collapse, i.e. only one AH (either AH^+ or AH^-) can occur. The AHs can thus cross only in an elliptic $E < 0$ region. At the neck of an L–T wormhole, where $2E = -1$, M is minimum, and t_B is maximum, the moment of maximum expansion is

$$R_{,t}{}^2 = 0 = \frac{2M}{R} - 1 \quad \rightarrow \quad R_{\max}(M_{\min}) = 2M. \tag{3.47}$$

At all other E values in an elliptic region $-1 < 2E < 0$, we find $R_{,t} = 0 \rightarrow R_{\max} = 2M/(-2E) > 2M$. Thus $R = 2M$ has two solutions – one in the expanding phase and one in the collapsing phase. So the AH^+ and AH^- can occur. The AHs meet at the neck maximum (set O).

To establish whether an AH is timelike, null or spacelike, we compare the slope of the AH^+ with the outgoing light ray (or of the AH^- with the incoming light ray), i.e. $\ell j = -1$:

$$B = \left. \frac{dt}{dr} \right|_{AH} \bigg/ \left. \frac{dt}{dr} \right|_{n} = \left(1 - \frac{2M_{,r}}{R_{,r}} \right), \tag{3.48}$$

but below, we will actually calculate, using (3.46),

$$\overline{B} = \left(\frac{R_{,r}}{M_{,r}} - 1 \right)_{AH\pm} = \frac{1+B}{1-B} = 1 - \ell\sqrt{1+2E} \, \frac{dt_{AH}}{dM}. \tag{3.49}$$

Now, since the conditions for no shell crossings (Hellaby and Lake, 1985) require $M_{,r} \geq 0$ where $R_{,r} > 0$ and vice-versa, we have

$$
\begin{aligned}
B_{\max} = 1, \ \ \overline{B} = \infty \quad &\rightarrow \ AH^+ \text{ outgoing null} \quad (M_{,r} = 0 = \rho), \\
-1 < B < 1, \ \ 0 < \overline{B} < \infty \quad &\rightarrow \ AH^+ \text{ spacelike} \quad (\text{for most } M_{,r}), \\
B = -1, \ \ \overline{B} = 0 \quad &\rightarrow \ AH^+ \text{ incoming null} \quad (M_{,r} = R_{,r}), \\
1 < -B < \infty, \ \ 0 < -\overline{B} < 1 \quad &\rightarrow \ AH^+ \text{ incoming timelike} \ (M_{,r} > R_{,r}),
\end{aligned}
\tag{3.50}
$$

so an outgoing timelike AH^+ is not possible. This means outgoing light rays that reach the AH^+ always fall inside it, except where $M_{,r} = 0$, in which case they move along it. The possibility that the $AH^\pm$ is timelike ($\overline{B} < 0$) holds if

$$\ell\frac{dt_{AH}}{dM} > \frac{1}{\sqrt{1+2E}}. \tag{3.51}$$

This means that the smaller E is, the steeper the locus of the AH must be to make it timelike.

The argument is similar for light rays at the AH^-, except that 'incoming' should be swapped for 'outgoing'. If incoming light rays reach the AH^-, they pass out of it or run along it.

Since this is true for every point on the AH^+ and AH^-, radial light rays in dense wormholes are more trapped and go even less far than in vacuum. In particular, if [1] $\rho > 0$ where $2E = -1$, light rays starting at O fall inside the AH^+ to the future and inside the AH^- to the past.

3.2.2 Event horizons

The event horizon is composed of the very last rays to reach future null infinity (EH^+) in each direction, or the very first ones to come in from past null infinity (EH^-). If we have vacuum ($M_{,r} = 0$) everywhere, then light rays travel along $R = 2M$, and the EHs coincide with the AHs. If there is matter, $M_{,r} > 0$, on any worldline, then the incoming light rays emerge from the AH^- and outgoing light rays fall into the AH^+ at that r value, and so the EHs split off from the AHs (see Hellaby, 1987).

3.2.3 Locating the AH^- during expansion in elliptic regions

We shall first consider the expansion phase of an elliptic model, where $0 \leq \eta \leq \pi$ and $E < 0$, so we have $\ell = +1$ and only the AH^- is present. Since $R = 2M$ on an AH, we have from (2.5a):

$$\cos \eta_{AH} = 1 + 4E. \tag{3.52}$$

Along a given worldline, the proper time of passing through the AH, counted from t_B, can be calculated from (2.11) with $R = 2M$ to be

$$t_{AH^-} - t_B = M \frac{\arccos(1 + 4E) - 2\sqrt{-2E(1 + 2E)}}{(-2E)^{3/2}}. \tag{3.53}$$

The function $F = t_{AH^-} - t_B$ of the argument $f = 2E$, defined in (3.53), has the following properties:

$$F(-1) = M\pi, \qquad F(0) = 4M/3,$$
$$\mathrm{d}F/\mathrm{d}f < 0 \qquad \text{for} \quad -1 < f < 0, \tag{3.54}$$

i.e. it is decreasing. These properties mean that the AH does not touch the Big Bang anywhere except at a centre, $M = 0$, even though (3.52) with $E = 0$ suggests otherwise.

Along all worldlines with $E > -1/2$ the dust particles emerge from the AH^- a finite time after the Big Bang ($\eta = 0$) and a finite time before maximum expansion ($\eta = \pi$). In order to find the slope of the AH^-, we differentiate (3.53) by M

---

[1] Since a regular minimum or maximum $R_{,r} = 0$ requires all of $M_{,r}$, $E_{,r}$ and $t_{B,r}$ to be locally zero, $\rho > 0 \Rightarrow M_{,rr}/R_{,rr} > 0$.

to obtain

$$\frac{\mathrm{d}t_{AH^-}}{\mathrm{d}M} = \left[\frac{1}{(-2E)^{3/2}} + \frac{3M}{(-2E)^{5/2}}\frac{\mathrm{d}E}{\mathrm{d}M}\right]\arccos(1 + 4E)$$

$$+ \frac{1}{E}\sqrt{1+2E} - \frac{M(3+2E)}{2E^2\sqrt{1+2E}}\frac{\mathrm{d}E}{\mathrm{d}M} + \frac{\mathrm{d}t_B}{\mathrm{d}M}. \tag{3.55}$$

This formula will prove useful later.

So, although (3.52) shows wordlines with larger E exit the AH^- at a later stage of evolution, this may not correspond to a later time t, or even to a longer time $(t - t_B)$ since the bang. It is not at all necessary that E is a monotonically decreasing function of r in an elliptic region; in general it can increase and decrease again any number of times.

3.2.4 AH⁻ near an origin in elliptic regions

On the AH^- in the neighbourhood of a regular origin, we obtain from (3.53), (2.16a) and (2.16c):

$$\lim_{M \to 0}(t_{AH^-} - t_B) = 0, \tag{3.56}$$

i.e. the AH^- touches the Big Bang set at the centre. Since $E = 0$ at $M = 0$ and $E < 0$ in the neighbourhood of the centre, it follows that $E_{,M}(0) \le 0$, and, via (3.52), that the dust particles with larger M (smaller E) exit the AH^- with larger values of η.

To show that the behaviour of the AH^- near a regular origin is not unique we take the following example for $E(M)$:

$$E = M^{2/3}(\gamma + \gamma_2 M^d), \quad d > 2/3, \quad \gamma \ne 0, \quad \gamma_2 \ne 0. \tag{3.57}$$

Putting (3.57) and (2.16d) in (3.55) we find,[2] neglecting powers of M that are necessarily positive,

$$\frac{\mathrm{d}t_{AH^-}}{\mathrm{d}M} \approx c\tau M^{c-1} + \frac{4}{3}, \tag{3.58}$$

so d plays no role here, and from (3.49) with $\ell = +1$ and (3.58) we see that

$$\overline{B} \approx -c\tau M^{c-1} - \frac{1}{3} + \cdots \tag{3.59}$$

---

[2] Since $M^{1/3}$ is a natural measure of proper radius near an origin, the slope of the AH^- with respect to $M^{1/3}$ is more meaningful geometrically, and it is zero:

$$\frac{\mathrm{d}t_{AH^-}}{\mathrm{d}M^{1/3}} \approx 3c\tau M^{c-1/3} + 4M^{2/3} + \cdots \xrightarrow[M \to 0]{} 0.$$

Away from the centre, the sign of $\mathrm{d}t_{AH^-}/\mathrm{d}M^{1/3}$ depends on whether or not $c > 1$, as $\tau < 0$ for no shell crossings (Hellaby and Lake, 1985).

The behaviour of the apparent horizon in the vicinity of the centre is not unique; each of the cases listed in (3.50) can occur, depending on whether $c > 1$. This nonuniqueness of behaviour is connected with the shell-focussing singularities that appear in some L–T models. Various studies[3] have shown that outgoing light rays may emerge from the central point of the Big Crunch and even reach infinity (see a simple example of this phenomenon presented by Plebański and Krasiński, 2006). Moreover, even though, in those cases, this central point (where the Big Crunch first forms) appears to be a single point in comoving coordinates, it is in fact a finite segment of a timelike or null[4] curve C_S in the Carter–Penrose diagram, see Eardley and Smarr (1979). At the Big Bang, we have the reverse – incoming light rays may reach the central point of the bang singularity. Any radial light ray emitted from the centre of symmetry that falls into the Big Crunch must first increase its R value, and then decrease, i.e. it must cross the AH$^+$ in between. In consequence, the AH$^+$ cannot touch the centre of the Big Crunch earlier than the C_S singularity does. Similarly, the AH$^-$ cannot touch the centre of the Big Bang later than the C_S singularity does.

3.2.5 AH$^+$ during collapse in elliptic regions

This can be obtained from the above by replacing $(t - t_B)$ with $(t_C - t)$, η with $(2\pi - \eta)$, flipping the signs of ℓ and j, and swopping 'incoming' for 'outgoing'. However, keeping t_B as our arbitrary function, we have $\pi \leq \eta \leq 2\pi$. Equation (3.52) still applies, but instead of (3.53) and (3.55), we now obtain

$$(t - t_B)_{AH^+} = M \frac{\pi + \arccos(-1 - 4E) + 2\sqrt{-2E(1 + 2E)}}{(-2E)^{3/2}}, \qquad (3.60)$$

$$\frac{dt_{AH^+}}{dM} = \left[\frac{1}{(-2E)^{3/2}} + \frac{3M}{(-2E)^{5/2}} \frac{dE}{dM} \right] [\pi + \arccos(-1 - 4E)]$$

$$- \frac{1}{E}\sqrt{1 + 2E} + \frac{M(3 + 2E)}{2E^2\sqrt{1 + 2E}} \frac{dE}{dM} + \frac{dt_B}{dM}, \qquad (3.61)$$

and (3.49) applies with $\ell = -1$. The special cases all follow in the same way. Near an origin, we find again nonunique behaviour, one of the possibilities being this time light rays escaping the crunch at a regular centre before the AH$^+$ forms. At a regular extremum where $E = -1/2$, $\eta_{AH^+} \to \pi$ and the AH$^+$ crosses the

---

[3] The studies were done for example in the following works: Eardley and Smarr (1979); Christodoulou (1984); Newman (1986); Newman and Joshi (1988); Gorini *et al.* (1989); Lemos (1991); Dwivedi and Joshi (1992); Joshi and Dwivedi (1993) and Joshi (1993, pp. 242–255).
[4] Several authors claim that this central singularity must be null, but we do not know why. In fact, table 1 of Hellaby and Lake (1988) shows its orientation depends on the path of approach, and indicates it has a lot of structure.

AH$^-$. In the parabolic limit, $\eta_{AH^+} \to 0$, $t_C - t_{AH^+} \to 4M/3$, and

$$\frac{\mathrm{d}t_{AH^+}}{\mathrm{d}M} \xrightarrow[E \to 0]{} -\frac{4}{3} + \frac{4M}{5}\frac{\mathrm{d}E}{\mathrm{d}M} + \frac{\mathrm{d}t_C}{\mathrm{d}M}, \tag{3.62a}$$

$$\overline{B}_{AH^+} \xrightarrow[E \to 0]{} -\frac{1}{3} + \frac{4M}{5}\frac{\mathrm{d}E}{\mathrm{d}M} + \frac{\mathrm{d}t_C}{\mathrm{d}M}, \tag{3.62b}$$

where t_C is given by (2.9) and (2.8).

3.2.6 AH$^-$ in the parabolic limit

A shell of parabolic worldlines occurs at the boundary between elliptic and hyperbolic regions, where $E \to 0$, but $E_{,r} \neq 0$ and[5] $M > 0$. From (3.52), (3.53), (2.8), (3.55) and (3.49) with $\ell = +1$, we see that

$$\eta_{AH^-} \xrightarrow[E \to 0]{} 0, \quad t_{AH^-} - t_B \xrightarrow[E \to 0]{} \frac{4M}{3}, \quad T \xrightarrow[E \to 0]{} \infty, \tag{3.63a}$$

$$\frac{\mathrm{d}t_{AH^-}}{\mathrm{d}M} \xrightarrow[E \to 0]{} \frac{4}{3} - \frac{4M}{5}\frac{\mathrm{d}E}{\mathrm{d}M} + \frac{\mathrm{d}t_B}{\mathrm{d}M}, \tag{3.63b}$$

$$\overline{B}_{AH^-} \xrightarrow[E \to 0]{} -\frac{1}{3} + \frac{4M}{5}\frac{\mathrm{d}E}{\mathrm{d}M} - \frac{\mathrm{d}t_B}{\mathrm{d}M}, \tag{3.63c}$$

so the AH$^-$ never touches the bang here, despite η being zero. The divergence of the worldline lifetime T shows that either the bang time or the crunch time recedes to infinity, as would be expected in a hyperbolic region. It is also possible that both these times diverge: an asymptotically flat model is achieved by letting $E \to 0$ as $r \to \infty$ (Hellaby, 1987), with both bang and crunch times diverging.[6] We also see the slope and the causal nature of the AH$^-$ are uncertain here. However, since by (2.8) $\mathrm{d}T/\mathrm{d}M = 2\pi/(-2E)^{3/2} + 6\pi M(\mathrm{d}E/\mathrm{d}M)/(-2E)^{5/2}$, at least one of $\mathrm{d}t_B/\mathrm{d}M$ and $\mathrm{d}t_C/\mathrm{d}M$ must diverge. Noting that $t_{B,r} \leq 0$ for no shell crossings, where t_B diverges, we find the AH$^-$ becomes incoming null towards the parabolic locus. Conversely, the AH$^+$ becomes outgoing null where t_C diverges.

3.2.7 Apparent horizons in parabolic regions

The corresponding results in an expanding $E = 0$, $E_{,r} = 0$ L–T model follow from Sec. 3.2.6 (or directly from (2.6) with $R = 2M$). In particular, (3.63b) and (3.63c) become

$$\frac{\mathrm{d}t_{AH^-}}{\mathrm{d}M} = \frac{4}{3} + \frac{\mathrm{d}t_B}{\mathrm{d}M}, \qquad \overline{B}_{AH^-} = -\frac{1}{3} - \frac{\mathrm{d}t_B}{\mathrm{d}M}. \tag{3.64}$$

Despite the no shell crossing condition $\mathrm{d}t_B/\mathrm{d}M < 0$, the AH$^-$ may still exhibit all possible behaviours of (3.50) with 'outgoing' and 'incoming' interchanged. In a collapsing parabolic model, time-reversed results apply.

---

[5] Elliptic and parabolic regions are not possible for $M = 0$.
[6] Another example of this kind will be discussed in Sec. 3.2.10.

3.2.8 Locating apparent horizons in hyperbolic regions

Using the same methods as for expanding elliptic regions, we find the behaviour of the AH in expanding hyperbolic regions is qualitatively the same. There is only one AH, no maximum expansion, and loci where $E = -1/2$ are not possible. But the results for origins and for the parabolic limit both carry over, except that a divergent dt_B/dM is not expected. Collapsing hyperbolic regions are essentially like collapsing elliptic regions.

3.2.9 Locating the event horizon

Locating the future EH requires integrating (3.41) backward in time from the point where it intersects the AH^+. But this point lies 'at infinity', which is an impossible location for the initial point in any numerical calculation. Analytical integration of (3.41) is in general hopeless, in consequence of the complicated form of the function $R(t, r)$. For particular models, the EH^+ may be located numerically using a compactified coordinate representation of the spacetime, see an example in Sec. 3.2.10.

3.2.10 An illustration – a simple recollapsing model

We shall illustrate several properties of the L–T model and of its apparent horizons using a simple example. In the present subsection, we will use this model with unrealistic parameter values, chosen so that all the figures are easily readable. Later, in Sec. 4.2, we will use the same model for modelling a galaxy with a black hole at the centre, with parameters chosen to fit observational data.

Definition

We take an $E < 0$ L–T model with a regular centre, whose Big Bang function $t_B(M)$ is

$$t_B(M) = -bM^2 + t_{B0}, \tag{3.65}$$

and whose Big Crunch function is

$$t_C(M) = aM^3 + T_0 + t_{B0}, \tag{3.66}$$

where the parameter T_0 is the lifetime of the central worldline $M = 0$. The numerical values of the parameters used in the figures will be $a = 2 \cdot 10^4$, $b = 200$, $t_{B0} = 5$, $T_0 = 0.05$. Since $t = t_C$ at $\eta = 2\pi$, we find from (2.5):

$$E(M) = -\frac{1}{2}\left(\frac{2\pi M}{t_C - t_B}\right)^{2/3} = -\left(\frac{\pi^2}{2}\right)^{1/3} \frac{M^{2/3}}{(aM^3 + bM^2 + T_0)^{2/3}}. \tag{3.67}$$

As $M \to \infty$, we have $t_B \to -\infty$, $t_C \to +\infty$ and $E \to 0$. Hence, the space contains infinite mass and has infinite volume. Unlike in Friedmann models, positive space

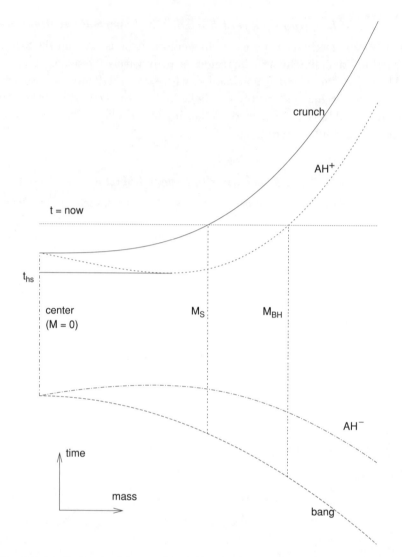

Fig. 3.1. Evolution leading to a black hole in the $E < 0$ L–T model of (3.65)–(3.66). The final state is defined at the instant $t = t_2 =$ now. Worldlines of dust particles are vertical straight lines, each has a constant mass-coordinate. Intersections of the line $t = t_2$ with the lines representing the Big Crunch and the future apparent horizon determine the masses M_S and M_{BH}, respectively.

curvature does not imply finite volume; this has been known since long ago (Bonnor, 1985; Hellaby and Lake, 1985).

Description

The main features of this model are shown in Fig. 3.1. The AH^+ first appears not at the centre, but at a finite distance from the centre, where the function $t_{AH^+}(M)$ has its minimum, and at a time $t_{hs} < t_C(0)$.

At all times after the crunch first forms, $t > t_C(0)$, the mass M_S already swallowed up by the singularity is necessarily smaller than the mass M_{BH} that has disappeared into the AH$^+$. For an astronomical object in which the black hole already exists, the time $t = $ now must be taken after t_{hs} and in Fig. 3.1 it is also greater than $t_C(0)$. The mass M_S cannot even be estimated by astronomical methods.

Figure 3.2 shows the 'topographic map' of the surface $R(t, M)$. It contains contours of constant R (the thinner curves) inscribed into Fig. 3.1. (The other curves are outgoing radial null geodesics, and we shall discuss them further below.)

It is evident from Fig. 3.2 that the AH$^+$ is spacelike everywhere except possibly in a neighbourhood of the central line $M = 0$ and at future null infinity. (The same is true for the AH$^-$.) From (3.67) and (3.65) we have $d = 2$ in (3.57) and $c = 2$ in (3.59), so we find $\overline{B}_{AH^-} \to 1$ at the origin. Similarly by (3.66) we have $c = 3$, and $\overline{B}_{AH^+} \to 1$. Therefore the AH$^\pm$ are both spacelike at the origin too.

Figure 3.2 also shows several outgoing radial null geodesics. Each geodesic has a vertical tangent at the centre. This is a consequence of using M as the radial coordinate. Since $dt/dM = \partial R/\partial M/\sqrt{1 + 2E}$ on each geodesic and $R \propto M^{1/3}$ close to the centre, so $dt/dM \propto M^{-2/3}$ and $dt/dM \to \infty$ as $M \to 0$. Each geodesic proceeds to higher values of R before it meets the apparent horizon AH$^+$. At the AH$^+$, it is tangent to an $R = $ const contour, then proceeds toward smaller R values. The future event horizon consists of those radial null geodesics that approach the AH$^+$ asymptotically. It will be shown to lie between the geodesics no 5 and 6 in the figure, counted from the lower right corner (see below).

Geodesic no 5 from the lower right emanates from the centre $M = 0$, where the Big Bang function has a local maximum. The tangent to the geodesic is horizontal there. This means that the observer receiving it sees the light infinitely redshifted, as in the Friedmann models. Geodesics to the right of this one all begin with a vertical tangent, which implies an infinite blueshift. These observations about redshift and blueshift were first made by Szekeres (1980) (see also Plebański and Krasiński, 2006). Likewise, the geodesics meet the Big Crunch with their tangents vertical.

By the time the crunch forms at $t = t_C(0)$, the future apparent horizon already exists (see Figs. 3.1 and 3.2). The shells of progressively greater values of M first go through the AH, and then hit the singularity at $t = t_C(M)$. At the time $t = t_2$, the singularity has already accumulated the mass M_S, while the mass hidden inside the apparent horizon at the same time is $M_{BH} > M_S$. Both of them grow with time, but at fixed t_2 they are constants. From the definitions of M_S and M_{BH} it follows that

$$t_2 = t_C(M_S) = t_{AH^+}(M_{BH}). \tag{3.68}$$

Location of the event horizon

Now we shall discuss the location of the event horizon in the spacetime model considered here. It will follow that, even though the model has a rather simple

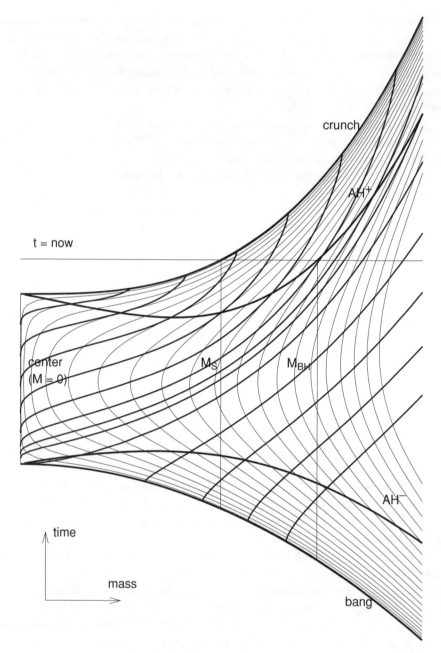

Fig. 3.2. Contours of constant R-value (thinner lines) and outgoing radial null geodesics inscribed into the spacetime diagram of Fig. 3.1. The R-values on consecutive contours differ by a uniform amount.

geometry, this is quite a complicated task that requires complete knowledge of the whole spacetime, including null infinity. Hence, in a real Universe, where our knowledge is limited to a relatively small neighbourhood of our past light cone and our past worldline, and the knowledge is mostly incomplete and imprecise, the event horizon simply cannot be located by astronomical observations.

The future event horizon is formed by those null geodesics that fall into the future apparent horizon 'as late as possible', i.e. approach it asymptotically. Hence, in order to locate the event horizon, we must issue null geodesics backward in time from the 'future end point' of the AH^+. This cannot be done in the (M, t) coordinates used so far because the spacetime and the AH^+ are infinite. Hence, we must first compactify the spacetime. The most convenient compactification for considering null geodesics at a null infinity is, theoretically, a Penrose transform because it maps the null infinities into finite sets. However, in order to find a Penrose transformation, one must first choose null coordinates, and in the L–T model this has so far proved to be an impossible task, see Stoeger *et al.* (1992) and Hellaby (1996a). Hence, we will use a less convenient compactification that will squeeze the null infinities into single points in the 2-dimensional (time–radius) spacetime diagram:

$$M = \tan(\mu), \qquad t = \tan(\tau). \tag{3.69}$$

In these coordinates, the $\mathbb{R}^1 \times \mathbb{R}^1_+$ space of Figs. 3.1–3.2 becomes the finite $[-\pi/2, \pi/2] \times [0, \pi/2]$ rectangle, see Fig. 3.3. The upper curve in the figure is the Big Crunch singularity; the future apparent horizon runs so close to it that it seems to coincide with it.[7] The horizontal line is the $\tau = $ now line. The lower curve is the Big Bang and the past apparent horizon, again running one on top of the other. The point on the τ-axis where the three lines meet is the image of the $M = 0$ line of Fig. 3.1, squeezed here into a point because of the scale of this figure.

The theoretical method to locate the future event horizon in Fig. 3.3 would now be to run a radial null geodesic backward in time from the point $(\mu, \tau) = (\pi/2, \pi/2)$, i.e. from the image of the future end of the AH^+. However, in the (μ, τ)-coordinates, for most of the range of μ the AH^+ runs so close to the crunch singularity, and the geodesics intersecting the AH^+ are so nearly tangent to the AH^+, that numerical instabilities crash any such geodesic into the singularity instantly, even if the initial point is chosen well away from $(\pi/2, \pi/2)$. This happens all the way down to $\mu = 0.5$ at single precision and all the way down to $\mu = 1.1$ at double precision. We did succeed, with double precision, only at $\mu = 1.0$, and a null geodesic could be traced from there to the centre at $\mu = 0$.

---

[7] As can be verified from (3.60), the time-difference between the crunch and the AH^+ goes to infinity when $M \to \infty$. However, the ratio of this time-difference to the crunch time goes to zero, which explains why the two curves in Fig. 3.3 meet at the image of the infinity. The same is true for the Big Bang and the AH^-.

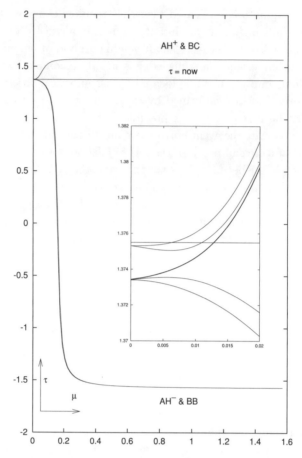

Fig. 3.3. The spacetime diagram of Fig. 3.1 compactified according to (3.69). The whole region shown in Fig. 3.1 is squeezed into the point where the three lines meet at the τ-axis. The worldlines of the dust source would still be vertical straight lines here. The upper curve is the future apparent horizon and the Big Crunch singularity, they seem to coincide at the scale of this picture. The lower curve is the past apparent horizon and the Big Bang singularity, again coinciding only at this resolution. The horizontal straight line is the time $\tau =$ now. Inset: a closeup view of the image (in the coordinates (μ, τ)) of the region shown in Fig. 3.1. The thicker line is the event horizon. It does not really hit the central point of the Big Bang; the apparent coincidence is just an artefact of the scale. More explanation in the text.

At the scale of Fig. 3.3, this whole geodesic seems to coincide with the crunch and the AH^+. However, it is well visible if one closes in on the image of the area shown in Fig. 3.1; the closeup is shown in the inset.

Actually, to make sure that we located the event horizon with an acceptable precision, we ran three different null geodesics backward in time; one from the point $\mu = 1.0, \tau = \tau_0 \stackrel{\text{def}}{=} \tau_{AH^+}(1.0)$ right on the apparent horizon, another one from $\mu = 1.0, \tau = \tau_1$, where τ_1 was in the middle between the AH^+ and

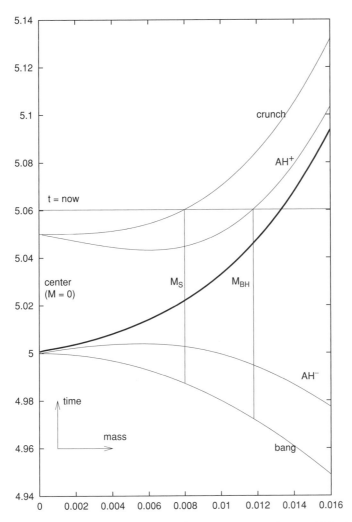

Fig. 3.4. The event horizon (thicker line) drawn into the frame of Fig. 3.1. Its intersection with the $M = 0$ axis does not coincide with the central point of the Big Bang, this is only an illusion created by the scale.

the crunch singularity, and the third one from $\mu = 1.0, \tau = \tau_2$, where τ_2 was below the AH$^+$, with the time-difference $\tau_0 - \tau_2 = \tau_1 - \tau_0$. All three ran so close to each other that they actually coalesced along the way and reached the centre as one curve.

Finally, the event horizon had to be transformed back to the (M, t)-coordinates and written into the frame of Figs. 3.1–3.2. This is done in Fig. 3.4. As stated earlier, the event horizon is located between the geodesics no 5 and 6 from the right in Fig. 3.2. By accident (caused by our choice of numerical values), the EH hits the centre very close to the central point of the Big Bang, but does not coincide with it.

This whole construction should make it evident that there is no chance to locate the event horizon by astronomical observations, even approximately. It only makes sense, in the observational context, to speak about an upper limit on the mass inside the *apparent horizon*. This is why we identified the observed mass of a black hole with the AH in Sec. 4.2.5.[8]

3.3 Null geodesics in the quasi-spherical Szekeres model

The geodesic equations $k^\alpha{}_{;\beta}k^\beta = 0$ (where $k^\alpha = \mathrm{d}x^\alpha/\mathrm{d}s$) in the quasi-spherical Szekeres model, are (see Hellaby and Krasiński, 2002)

$\alpha = 0$:

$$\frac{\mathrm{d}^2 t}{\mathrm{d}s^2} + \frac{\Phi_{,tr} - \Phi_{,t}\mathcal{E}_{,r}/\mathcal{E}}{1 - k}(\Phi_{,r} - \Phi\mathcal{E}_{,r}/\mathcal{E})\left(\frac{\mathrm{d}r}{\mathrm{d}s}\right)^2$$
$$+ \frac{\Phi\Phi_{,t}}{\mathcal{E}^2}\left[\left(\frac{\mathrm{d}x}{\mathrm{d}s}\right)^2 + \left(\frac{\mathrm{d}y}{\mathrm{d}s}\right)^2\right] = 0, \tag{3.70}$$

$\alpha = 1$:

$$\frac{\mathrm{d}^2 r}{\mathrm{d}s^2} + 2\frac{\Phi_{,tr} - \Phi_{,t}\mathcal{E}_{,r}/\mathcal{E}}{\Phi_{,r} - \Phi\mathcal{E}_{,r}/\mathcal{E}}\frac{\mathrm{d}t}{\mathrm{d}s}\frac{\mathrm{d}r}{\mathrm{d}s}$$
$$+ \left(\frac{\Phi_{,rr} - \Phi_{,r}\mathcal{E}_{,r}/\mathcal{E} - \Phi\mathcal{E}_{,rr}/\mathcal{E} + \Phi(\mathcal{E}_{,r}/\mathcal{E})^2}{\Phi_{,r} - \Phi\mathcal{E}_{,r}/\mathcal{E}} + \frac{1}{2}\frac{k_{,r}}{1 - k}\right)\left(\frac{\mathrm{d}r}{\mathrm{d}s}\right)^2$$
$$+ 2\frac{\Phi}{\mathcal{E}^2}\frac{\mathcal{E}_{,r}\mathcal{E}_{,x} - \mathcal{E}\mathcal{E}_{,xr}}{\Phi_{,r} - \Phi\mathcal{E}_{,r}/\mathcal{E}}\frac{\mathrm{d}r}{\mathrm{d}s}\frac{\mathrm{d}x}{\mathrm{d}s} + 2\frac{\Phi}{\mathcal{E}^2}\frac{(\mathcal{E}_{,r}\mathcal{E}_{,y} - \mathcal{E}\mathcal{E}_{,yr})}{\Phi_{,r} - \Phi\mathcal{E}_{,r}/\mathcal{E}}\frac{\mathrm{d}r}{\mathrm{d}s}\frac{\mathrm{d}y}{\mathrm{d}s}$$
$$- \frac{\Phi}{\mathcal{E}^2}\frac{1 - k}{\Phi_{,r} - \Phi\mathcal{E}_{,r}/\mathcal{E}}\left[\left(\frac{\mathrm{d}x}{\mathrm{d}s}\right)^2 + \left(\frac{\mathrm{d}y}{\mathrm{d}s}\right)^2\right] = 0, \tag{3.71}$$

$\alpha = 2$:

$$\frac{\mathrm{d}^2 x}{\mathrm{d}s^2} + 2\frac{\Phi_{,t}}{\Phi}\frac{\mathrm{d}t}{\mathrm{d}s}\frac{\mathrm{d}x}{\mathrm{d}s} - \left(\frac{1}{\Phi}\frac{\Phi_{,r} - \Phi\mathcal{E}_{,r}/\mathcal{E}}{1 - k}(\mathcal{E}_{,r}\mathcal{E}_{,x} - \mathcal{E}\mathcal{E}_{,xr})\right)\left(\frac{\mathrm{d}r}{\mathrm{d}s}\right)^2$$
$$+ 2\left(\frac{\Phi_{,r}}{\Phi} - \frac{\mathcal{E}_{,r}}{\mathcal{E}}\right)\frac{\mathrm{d}r}{\mathrm{d}s}\frac{\mathrm{d}x}{\mathrm{d}s} - \frac{\mathcal{E}_{,x}}{\mathcal{E}}\left(\frac{\mathrm{d}x}{\mathrm{d}s}\right)^2 - 2\frac{\mathcal{E}_{,y}}{\mathcal{E}}\frac{\mathrm{d}x}{\mathrm{d}s}\frac{\mathrm{d}y}{\mathrm{d}s}$$
$$+ \frac{\mathcal{E}_{,x}}{\mathcal{E}}\left(\frac{\mathrm{d}y}{\mathrm{d}s}\right)^2 = 0, \tag{3.72}$$

---

[8] Note that this model, in which the event horizon has been found, will actually be used only for the interior of our galaxy model – i.e. for that part of the galaxy that is invisible to outside observers because of its proximity to the black hole horizon. The model of the visible part of the galaxy will be different, and to locate the event horizon there, the whole construction (integrating a null geodesic backward in time from future null infinity) would have to be repeated. But, of course, the model of a single galaxy does not extend to infinity. Hence, it hardly makes sense to even speak of an event horizon in this context.

$\alpha = 3$:

$$\frac{d^2 y}{ds^2} + 2\frac{\Phi_{,t}}{\Phi}\frac{dt}{ds}\frac{dy}{ds} - \left(\frac{1}{\Phi}\frac{\Phi_{,r} - \Phi\mathcal{E}_{,r}/\mathcal{E}}{1-k}(\mathcal{E}_{,r}\,\mathcal{E}_{,y} - \mathcal{E}\mathcal{E}_{,yr})\right)\left(\frac{dr}{ds}\right)^2$$

$$+ 2\left(\frac{\Phi_{,r}}{\Phi} - \frac{\mathcal{E}_{,r}}{\mathcal{E}}\right)\frac{dr}{ds}\frac{dy}{ds} + \frac{\mathcal{E}_{,y}}{\mathcal{E}}\left(\frac{dx}{ds}\right)^2 - 2\frac{\mathcal{E}_{,x}}{\mathcal{E}}\frac{dx}{ds}\frac{dy}{ds}$$

$$- \frac{\mathcal{E}_{,y}}{\mathcal{E}}\left(\frac{dy}{ds}\right)^2 = 0. \tag{3.73}$$

When studying light propagation in the quasi-spherical Szekeres model, these equations, together with the evolution equations and the condition $k_\alpha k^\alpha = 0$, must be solved. Apart from special cases which will be discussed in the next subsection, these equations do not reduce to a simpler form. Also, in the general case there is no simple formula for redshift – to find the frequency shift z the general formula $1 + z = k_e^0/k^0$ (where the subscript e refers to the emission instant) has to be employed.

3.3.1 Constant (x, y) geodesics

When studying light propagation in the Friedmann or L–T models, radial null geodesics are often considered. Along such a geodesic only the time and the radial coordinate change, while the other two spatial coordinates remain constant. However, in the Szekeres model, in general it is impossible to define a radial direction. For example, one cannot impose $dx = dy = 0$ on a solution of (3.70)–(3.73) with a general metric because this would imply $d^2x/ds^2 \neq 0 \neq d^2y/ds^2$, i.e. the condition $dx = dy = 0$ would not be preserved along the curve. A null geodesic with $dx = dy = 0$ all along (we will temporarily call it a 'constant-(x, y) geodesic', but see at the end of this subsection) can occur only in one very special case. This is not a special geodesic in a general Szekeres spacetime, but just one geodesic in one special Szekeres spacetime. We mention this rather simple subject for the record, because erroneous statements on this matter flashed through the literature. What we show below has in fact been already proved by Nolan and Debnath (2007), but we adapt their result to our notation.

As follows from (3.70)–(3.73) if initially $k^x = k^y = 0$, then the coordinates x and y will remain constant only if, along the whole geodesic

$$\Phi_{,r} = \Phi\mathcal{E}_{,r}/\mathcal{E}, \tag{3.74}$$

or

$$\mathcal{E}\mathcal{E}_{,xr} = \mathcal{E}_{,x}\,\mathcal{E}_{,r}, \qquad \mathcal{E}\mathcal{E}_{,yr} = \mathcal{E}_{,y}\,\mathcal{E}_{,r}. \tag{3.75}$$

The relation (3.74) holds only at a shell-crossing singularity, which must be eliminated in a physically acceptable model. Let us then focus on condition (3.75). In the quasi-spherical model $\mathcal{E} > 0$ for all r. From the definition of $\mathcal{E}$,

(2.150), we have

$$\mathcal{E}_{,x} = \frac{x - P}{S}, \qquad \mathcal{E}_{,xr} = -\frac{P_{,r}}{S} - \frac{S_{,r}}{S}\mathcal{E}_{,x}, \qquad (3.76)$$

$$\mathcal{E}_{,y} = \frac{y - Q}{S}, \qquad \mathcal{E}_{,yr} = -\frac{Q_{,r}}{S} - \frac{S_{,r}}{S}\mathcal{E}_{,y}. \qquad (3.77)$$

Note that (3.75) are not functional relations – they define special values of x and y (and special forms of $P(r)$, $Q(r)$ and $S(r)$) at which the equalities should hold along a whole null geodesic. Thus, statements like $\mathcal{E}_{,x} = 0$ or $\mathcal{E}_{,y} = 0$ below should also be understood as defining special locations and forms of functions, and not as functional relations.

Remembering this, we have to consider the following cases:

1. $\mathcal{E}_{,x} \neq 0 \neq \mathcal{E}_{,y}$.
   Then, from (3.75)

$$\frac{\mathcal{E}_{,xr}}{\mathcal{E}_{,x}} = \frac{\mathcal{E}_{,yr}}{\mathcal{E}_{,y}} \iff \frac{P_{,r}}{x - P} = \frac{Q_{,r}}{y - Q}. \qquad (3.78)$$

This should hold at a constant $x = x_0$ and a constant $y = y_0$ along the whole geodesic, i.e. at every r. This equation is fulfilled for any P and any Q at $(x_0, y_0) \to (\pm\infty, \pm\infty)$ (in the stereographic coordinates the point $x_0 = +\infty$ coincides with $x_0 = -\infty$, and the same is true for y_0). However, $(x_0, y_0) = (\infty, \infty)$ is a singular point of the geodesic equations (3.70)–(3.73) and of the metric (2.150), so this case must be investigated separately. We will discuss it after dealing with the other cases.

In seeking for other solutions, we can integrate (3.78) with respect to r. One special solution of this equation is $P_{,r} = Q_{,r} = 0$; we will come back to this case below. The general integral is

$$Q = y_0 + C_0(P - x_0) \qquad (3.79)$$

along the geodesic, where C_0 is an arbitrary constant.

But with $\mathcal{E}_{,x} \neq 0 \neq \mathcal{E}_{,y}$, (3.75) also implies $\mathcal{E}_{,xr}/\mathcal{E}_{,x} = \mathcal{E}_{,r}/\mathcal{E}$, i.e.

$$\frac{P_{,r}}{x_0 - P} = \frac{2\left[(x_0 - P)P_{,r} + (y_0 - Q)Q_{,r}\right] - 2SS_{,r}}{(x_0 - P)^2 + (y_0 - Q)^2 + S^2}. \qquad (3.80)$$

This is again fulfilled for any P and any Q at $(x_0, y_0) \to (\pm\infty, \pm\infty)$, which case will be discussed later. If $P_{,r} = 0$ in (3.80), then $Q_{,r} = 0$ from (3.79), and then the above implies $S_{,r} = 0$, which reduces the Szekeres model to Lemaître–Tolman.

Apart from this obvious special case, with $P_{,r} \neq 0$, (3.80) is easily integrated with the result:

$$S^2 = C_1(x_0 - P) - (C_0{}^2 + 1)(x_0 - P)^2, \qquad (3.81)$$

where C_1 is another arbitrary constant [and (3.79) was used to eliminate $(y_0 - Q)$]. Note that $C_1 \neq 0$.

2. $\mathcal{E}_{,y} = 0 \neq \mathcal{E}_{,x}$.

   With $\mathcal{E}_{,y} = (y - Q)/S = 0$ we have $y_0 = Q$ along the whole geodesic, i.e.
   $Q_{,r} = 0$. Then the second of (3.77) shows that also $\mathcal{E}_{,yr} = 0$ along this geodesic,
   and the second of (3.75) is fulfilled. The first of (3.75) then becomes a special
   case of (3.80):

$$\frac{P_{,r}}{x_0 - P} = \frac{2(x_0 - P)P_{,r} - 2SS_{,r}}{(x_0 - P)^2 + S^2}. \tag{3.82}$$

   This again has the special solutions $(x_0, y_0) = (\infty, \infty)$, to be discussed below,
   and $P_{,r} = S_{,r} = 0$, which is the L–T model. The general integral of (3.82) is
   the subcase $C_0 = 0$ of (3.81).
3. $\mathcal{E}_{,y} \neq 0 = \mathcal{E}_{,x}$.

   As is easy to predict, the result here follows from the previous case by simple
   interchanges of coordinates and functions. The general solution here is

$$P = x_0 \implies P_{,r} = 0, \qquad S^2 = C_2(y_0 - Q) - (y_0 - Q)^2. \tag{3.83}$$

4. $\mathcal{E}_{,y} = 0 = \mathcal{E}_{,x}$.

   This implies, via (3.76)–(3.77):

$$(x_0, y_0) = (P, Q) \implies P_{,r} = Q_{,r} = 0, \tag{3.84}$$

   and (3.75) is then fulfilled without any further conditions. With $P_{,r} = Q_{,r} = 0$
   the model becomes axially symmetric, and $(x_0, y_0) = (P, Q)$ is the axis of
   symmetry in each 3-space of constant t.

   Now it can be verified that all the four cases discussed above are equivalent to
each other under coordinate transformations. The general form of the Szekeres
metric (2.150) is preserved by the 2-dimensional Haantjes transformations (see
pp. 92–93 in Plebański and Krasiński, 2006). We combine the shift $(x, y) \to$
$(x - x_0, y - y_0)$ with a Haantjes transformation:

$$x = x_0 + \frac{x' + D_1\left(x'^2 + y'^2\right)}{T},$$

$$y = y_0 + \frac{y' + D_2\left(x'^2 + y'^2\right)}{T},$$

$$T \stackrel{\text{def}}{=} 1 + 2D_1 x' + 2D_2 y' + \left(D_1{}^2 + D_2{}^2\right)\left(x'^2 + y'^2\right), \tag{3.85}$$

where D_1 and D_2 are arbitrary constants. A characteristic property of this transformation is

$$(x - x_0)^2 + (y - y_0)^2 = \frac{x'^2 + y'^2}{T},$$

$$dx^2 + dy^2 = \frac{dx'^2 + dy'^2}{T^2}. \tag{3.86}$$

Applying the transformation (3.85) to the Szekeres metric whose P, Q and S obey (3.79) and (3.81) we see that if

$$D_1 = -1/C_1, \qquad D_2 = -C_0/C_1, \qquad (3.87)$$

then the new $\mathcal{E}$ will be of the form $\left(x'^2 + y'^2\right)/(2\tilde{S}) + \tilde{S}/2$, where

$$\tilde{S} = C_1 \sqrt{\frac{x_0 - P}{C_1 - \left(C_0{}^2 + 1\right)(x_0 - P)}}, \qquad (3.88)$$

i.e. the resulting metric will be axially symmetric.

Since (3.82) follows from (3.79)–(3.81) as the subcase $C_0 = 0$, this is obviously axially symmetric, too.

For (3.83) an explicitly axially symmetric metric is obtained by taking $D_1 = 0$ and $D_2 = -1/C_2$ in (3.85).

The image of the position of the geodesic at $(x, y) = (x_0, y_0)$ is, in each of the above cases, the set $x' = y' = 0$, which is evidently the axis of symmetry in the (x', y')-plane.

Now we deal with the point $(x_0, y_0) = (\infty, \infty)$. To check the behaviour of geodesics passing through this point we transform (2.150) by

$$(x, y) = \frac{(x', y')}{x'^2 + y'^2}. \qquad (3.89)$$

The image of the surface $(x, y) = (\infty, \infty)$ under this transformation is the surface $(x', y') = (0, 0)$. Equation (3.89) transforms (2.150) into a Szekeres metric still of the form (2.150), but with $\mathcal{E}$ replaced by

$$\tilde{\mathcal{E}} = \frac{1}{2\tilde{S}}\left[\left(x' - \tilde{P}\right)^2 + \left(y' - \tilde{Q}\right)^2\right] + \frac{\tilde{S}}{2}, \qquad (3.90)$$

where

$$\left(\tilde{P}, \tilde{Q}, \tilde{S}\right) = \frac{(P, Q, S)}{P^2 + Q^2 + S^2}. \qquad (3.91)$$

We now verify whether a null geodesic on which initially $(x', y') = (0, 0)$ preserves $(x', y') = (0, 0)$ all along it. The investigation above suggests that this should be the case for every form of (P, Q, S). If this were true, then the relations (3.75) would now be identities obeyed by $\tilde{\mathcal{E}}$ at $x' = y' = 0$, for any functions $\left(\tilde{P}, \tilde{Q}, \tilde{S}\right)$.

However, this cannot be true: equations (3.75) applied to $\tilde{\mathcal{E}}(x', y')$ at $x' = y' = 0$ are of the same algebraic form as (3.75) applied to $\mathcal{E}(x, y)$ at $x = y = 0$, i.e. they will impose limitations on $\left(\tilde{P}, \tilde{Q}, \tilde{S}\right)$. We show just one example of this. Equations (3.75) applied to $\tilde{\mathcal{E}}(x', y')$ imply

$$\tilde{\mathcal{E}}_{,y'}\,\tilde{\mathcal{E}}_{,x'r} = \tilde{\mathcal{E}}_{,x'}\,\tilde{\mathcal{E}}_{,y'r}. \qquad (3.92)$$

Substituting for $\widetilde{\mathcal{E}}$ from (3.90) at $x' = y' = 0$ we get $\widetilde{P}_{,r}\,\widetilde{Q} = \widetilde{P}\widetilde{Q}_{,r}$. Apart from the special case $\widetilde{P}_{,r} = \widetilde{Q}_{,r} = 0$ this is integrated with the result $\widetilde{Q} = A\widetilde{P}$, where A is a constant. From (3.91), this is equivalent to $Q = AP$. This means that (3.92) is not identically obeyed at $x' = y' = 0$. Thus, apart from special forms of P, Q and S, a null geodesic on which initially $(x, y) = (\infty, \infty)$ is not a constant-(x, y) geodesic.

The final conclusion of our considerations is:

Theorem 3.1 *A constant-(x, y) null geodesic with constant x and y exists only in an axially symmetric Szekeres spacetime. Apart from the L–T limit, there is only one such geodesic – the one which stays on the axis of symmetry within each (x, y) surface. Depending on the coordinates, the functions P, Q and S in such a spacetime obey one of the sets of equations:*

(a) (3.79) and (3.81);
(b) the subcase $C_0 = 0$ of (3.79)–(3.81);
(c) (3.83);
(d) (3.84). $\square$

Although it is possible that there are other geodesics which have a fixed direction, for example geodesics along which x/y is constant, here we only considered geodesics which are axially directed. Let us call this kind of geodesics the axial geodesics.

3.3.2 The redshift formula for the axial geodesics

In the case of the axial geodesics one can set $dx = 0 = dy$ to obtain the following equation for a null geodesic:

$$\frac{dt}{dr} = -\frac{\Phi_{,r} - \Phi\mathcal{E}_{,r}/\mathcal{E}}{\sqrt{1-k}}. \tag{3.93}$$

We chose the $-$ sign, so the observer is at the origin. Let us choose the following parametrisation:

$$k^1 = -1, \qquad k^0 = \frac{\Phi_{,r} - \Phi\mathcal{E}_{,r}/\mathcal{E}}{\sqrt{1-k}}. \tag{3.94}$$

Since this in not an affine parametrisation, the parallel transport does not preserve the tangent vector. Thus k^α, after being parallel transported, becomes λk^α (where λ is a scalar coefficient – a function of the parameter s along the geodesic). In this case the geodesic equations are of the form (Plebański and Krasiński, 2006):

$$k^\alpha{}_{;\beta}k^\beta = -\frac{1}{\lambda}\frac{d\lambda}{ds}k^\alpha. \tag{3.95}$$

The above is equivalent to only one equation:

$$-\frac{1}{\lambda}\frac{d\lambda}{ds} = 2\frac{\Phi_{,tr} - \Phi_{,t}\,\mathcal{E}_{,r}\,/\mathcal{E}}{\sqrt{1-k}} - \frac{\Phi_{,rr} - \Phi_{,r}\,\mathcal{E}_{,r}\,/\mathcal{E} - \Phi\mathcal{E}_{,rr}\,/\mathcal{E} + \Phi(\mathcal{E}_{,r}\,/\mathcal{E})^2}{\Phi_{,r} - \Phi\mathcal{E}_{,r}\,/\mathcal{E}}$$

$$-\frac{1}{2}\frac{k_{,r}}{1-k}. \tag{3.96}$$

All the quantities above are evaluated on the geodesic, where t and r are connected with each other via (3.93). Thus we have

$$\Phi_{n,r} = (\Phi_{,t})_n\frac{dt}{dr} + (\Phi_{,r})_n,$$

$$(\Phi_{,r})_{n,r} = (\Phi_{,tr})_n\frac{dt}{dr} + (\Phi_{,rr})_n, \tag{3.97}$$

where the subscript n refers to quantities measured on the geodesic.

The second term on the right-hand side of (3.96) looks like a logarithmic derivative. However because of (3.97) we have

$$\frac{d\ln\left[(\Phi_{,r})_n - \Phi_n(\mathcal{E}_{,r}\,/\mathcal{E})_n\right]}{dr}$$

$$= \frac{(\Phi_{,r})_{n,r} - \Phi_{n,r}\,(\mathcal{E}_{,r}\,/\mathcal{E})_n - \Phi_n(\mathcal{E}_{,rr}\,/\mathcal{E})_n + \Phi_n(\mathcal{E}_{,r}\,/\mathcal{E})_n^2}{(\Phi_{,r})_n - \Phi_n(\mathcal{E}_{,r}\,/\mathcal{E})_n}$$

$$= \frac{(\Phi_{,rr})_n - (\Phi_{,r})_n(\mathcal{E}_{,r}\,/\mathcal{E})_n - \Phi_n(\mathcal{E}_{,rr}\,/\mathcal{E})_n + \Phi_n(\mathcal{E}_{,r}\,/\mathcal{E})_n^2}{(\Phi_{,r})_n - \Phi_n(\mathcal{E}_{,r}\,/\mathcal{E})_n}$$

$$+ \frac{(\Phi_{,tr})_n - (\Phi_{,t})_n(\mathcal{E}_{,r}\,/\mathcal{E})_n}{(\Phi_{,r})_n - \Phi_n(\mathcal{E}_{,r}\,/\mathcal{E})_n}\frac{dt}{dr}. \tag{3.98}$$

Using the above relation we can integrate equation (3.96):

$$\lambda = C\frac{\sqrt{1-k_n}}{(\Phi_{,r})_n - \Phi_n(\mathcal{E}_{,r}\,/\mathcal{E})_n}\,\exp\left(\int dr\frac{(\Phi_{,tr})_n - (\Phi_{,t})_n(\mathcal{E}_{,r}\,/\mathcal{E})_n}{\sqrt{1-k_n}}\right). \tag{3.99}$$

Now we can easily find that k^α in the affine parametrisation is given by $\widetilde{k}^\alpha = (\lambda/C)k^\alpha$. Using the general formula for redshift, $1 + z = \widetilde{k}_e^0/\widetilde{k}_o^0$, we can write an easy-to-use relation for redshift:

$$\ln(1+z) = l\int_{r_e}^{r_o} dr\frac{(\Phi_{,tr})_n - (\Phi_{,t})_n(\mathcal{E}_{,r}\,/\mathcal{E})_n}{\sqrt{1-k_n}}, \tag{3.100}$$

where $l = 1$ for $r_e < r_o$ and $l = -1$ for $r_e > r_o$. This relation is valid only on the axial geodesics, thus one of the conditions listed in Sec. 3.3.1 must hold. Under such conditions light propagation studies become much simpler. One needs to solve only one equation, i.e. (3.93), and the redshift can be directly calculated from (3.100).

3.4 Junction of null geodesics in the Szekeres Swiss-cheese model

The Swiss-cheese models are often used to take into account the inhomogeneous matter distribution observed in our Universe. However, when employing such models to the analysis of astronomical observations, junction conditions must be handled properly. Whereas the junction of two spacetimes is well established (the first and the second fundamental forms, $^3g_{ij}$ and K_{ij}, must remain continuous) the junction of null geodesics may introduce some complications, so the subject is considered in this section.

For a curve $x^\gamma(\lambda)$ with tangent vector $X^\sigma = \mathrm{d}x^\sigma/\mathrm{d}\lambda$ crossing the junction Σ, the obvious junction conditions are the continuity of its position and its tangent vector, but expressed in a form that makes them invariant with respect to the transformations of the 4D coordinates, x^γ. The two manifolds on either side of Σ are labelled '+' and '−', and the surface Σ is defined on each side by

$$x^\gamma_\pm = x^\gamma_\pm(\xi^k),$$

where ξ^k, $k = 1, 2, 3$ are surface coordinates. For the curve to be of class C^0 we need to ensure the curve intersects Σ at the same intrinsic coordinate point, i.e.:

$$[\xi^k(x^\gamma)] = \xi^k(x^\gamma_+(\lambda^+_\Sigma)) - \xi^k(x^\gamma_-(\lambda^-_\Sigma)) = 0. \tag{3.101}$$

For the curve to be of class C^1, the components of the tangent vector, projected normal to and tangent to Σ, must be continuous:

$$\left[X^\beta n_\beta\right] = \left.\frac{\mathrm{d}x^\beta_+}{\mathrm{d}\lambda^+} n^+_\beta\right|_\Sigma - \left.\frac{\mathrm{d}x^\beta_-}{\mathrm{d}\lambda^-} n^-_\beta\right|_\Sigma = 0,$$

$$\left[X_\alpha e_i{}^\alpha\right] = \left.\frac{\mathrm{d}x^\alpha_+}{\mathrm{d}\lambda^+} g_{\alpha\beta}\, e_i{}^\beta\right|_\Sigma - \left.\frac{\mathrm{d}x^\alpha_-}{\mathrm{d}\lambda^-} g_{\alpha\beta}\, e_i{}^\beta\right|_\Sigma = 0, \tag{3.102}$$

where n^α is the unit normal to Σ, $e_i{}^\alpha = \partial x^\alpha/\partial\xi^i$ are three orthogonal unit tangent vectors to Σ and square brackets indicate the jump in a quantity across the junction. In fact, the curve should still be C^1 even if we re-parametrise the curve on one side only, $\overline{\lambda} = \overline{\lambda}(\lambda)$. This would re-scale all components of the tangent vector by the same factor. Thus it is really the ratio of components that matters:

$$\left[(X_\alpha e_i{}^\alpha)\Big/(X^\beta n_\beta)\right] = 0. \tag{3.103}$$

If the curve is null, then it suffices to set

$$[\ell_\alpha e_i{}^\alpha] = \left[\frac{\mathrm{d}x^\alpha}{\mathrm{d}\lambda} g_{\alpha\beta} e_i{}^\beta\right] = 0, \tag{3.104}$$

and use $\ell^\alpha \ell_\alpha = 0$ to fix the normal component.

For our Swiss-cheese model, let us consider Szekeres spheres matched into a Friedmann background, and let us imagine a light ray passes out of one

Szekeres inhomogeneity straight into another one, where the boundary spheres touch. We can compress the double matching into a single junction calculation – Szekeres to Szekeres. Let us denote the null vector in the Szekeres model by $\ell^\alpha = (\ell^t, \ell^r, \ell^x, \ell^y)$. Since it is null, one component is set, let us say ℓ^r:

$$\ell^r = \frac{\sqrt{(\varepsilon - k)\left[(\ell^t)^2 - (\Phi^2/\mathcal{E}^2)\{(\ell^x)^2 + (\ell^y)^2\}\right]}}{\Phi_{,r} - \Phi\mathcal{E}_{,r}/\mathcal{E}}. \tag{3.105}$$

A surface of constant r, with surface coordinates (t, x, y), has

$$n_\alpha = \left(0, \frac{\Phi_{,r} - \Phi\mathcal{E}_{,r}/\mathcal{E}}{\sqrt{\varepsilon - k}}, 0, 0\right),$$

$$e_i{}^\alpha = \begin{pmatrix} 1 & 0 & 0 \\ 0 & 0 & 0 \\ 0 & 1 & 0 \\ 0 & 0 & 1 \end{pmatrix},$$

$$^3 g_{ij} = \begin{pmatrix} 1 & 0 & 0 \\ 0 & -\Phi^2/\mathcal{E}^2 & 0 \\ 0 & 0 & -\Phi^2/\mathcal{E}^2 \end{pmatrix},$$

$$K_{ij} = \begin{pmatrix} 0 & 0 & 0 \\ 0 & -\Phi\sqrt{\varepsilon - k}/\mathcal{E}^2 & 0 \\ 0 & 0 & -\Phi\sqrt{\varepsilon - k}/\mathcal{E}^2 \end{pmatrix}. \tag{3.106}$$

Thus

$$\ell_\alpha e_i{}^\alpha = \left(\ell^t, -\Phi^2 \ell^x/\mathcal{E}^2, -\Phi^2 \ell^y/\mathcal{E}^2\right). \tag{3.107}$$

The components of ℓ^α in the orthonormal tetrad $(n^\alpha, e_i{}^\alpha)$ are

$$\ell^{(\alpha)} = \left(\ell^t, \frac{\Phi_{,r} - \Phi\mathcal{E}_{,r}/\mathcal{E}}{\sqrt{\varepsilon - k}}\ell^r, \frac{\Phi}{\mathcal{E}}\ell^x, \frac{\Phi}{\mathcal{E}}\ell^y\right). \tag{3.108}$$

The matching requires that three of the following equations are obeyed:

$$\ell^t\big|_{\Sigma+} = \ell^t\big|_{\Sigma-},$$

$$\frac{\Phi_{,r} - \Phi\mathcal{E}_{,r}/\mathcal{E}}{\sqrt{\varepsilon - k}}\ell^r\bigg|_{\Sigma+} = \frac{\Phi_{,r} - \Phi\mathcal{E}_{,r}/\mathcal{E}}{\sqrt{\varepsilon - k}}\ell^r\bigg|_{\Sigma-},$$

$$\frac{\Phi}{\mathcal{E}}\ell^x\bigg|_{\Sigma+} = \frac{\Phi}{\mathcal{E}}\ell^x\bigg|_{\Sigma-},$$

$$\frac{\Phi}{\mathcal{E}}\ell^y\bigg|_{\Sigma+} = \frac{\Phi}{\mathcal{E}}\ell^y\bigg|_{\Sigma-}, \tag{3.109}$$

while the fourth one (say the r equation) follows from the null condition.

Now let us notice that when a Szekeres sphere is matched to a Friedmann background, its orientation is completely undetermined. If we fix the point where a light ray exits one Szekeres sphere and where it enters another, then one can still be rotated around the common normal direction relative to the other. This is equivalent to rotating the tangential component of the null vector. Thus we only need to match up the time component and the tangential component:

$$[\ell^t] = 0 = \left[\frac{\Phi^2}{\mathcal{E}^2}\left((\ell^x)^2 + (\ell^y)^2\right)\right]. \tag{3.110}$$

3.5 The apparent horizons in the Szekeres model

An apparent horizon (AH) is the boundary of the region of trapped surfaces. A trapped surface is a closed surface on which both the inward- and outward-directed null *geodesics* converge (i.e. have a negative expansion scalar). Thus, for a trapped surface Σ, if k^μ is any field of vectors tangent to null geodesics that intersect Σ, then

$$k^\mu{}_{;\mu} < 0 \qquad \text{on } \Sigma. \tag{3.111}$$

Consequently, on an apparent horizon:

$$k^\mu{}_{;\mu} = 0. \tag{3.112}$$

In the above, k^μ is null *and geodesic*. Proceeding from this definition, Szekeres (1975b) found that in the quasi-spherical Szekeres model the apparent horizon is given by the same equation as in the L–T model:

$$\Phi = 2M. \tag{3.113}$$

On the other hand, Hellaby and Krasiński (2002), while investigating the possibility of travelling through a Szekeres wormhole, defined a related notion, which they also called apparent horizon. Namely, they considered surfaces from which even a fastest-moving object could not escape outwards. Thus, they considered the analogue of (3.112), but for *non-geodesic* null fields. The reasoning was that where a geodesic ray would already be forced inward, an accelerating ray might still escape farther before being turned back. To avoid confusion, we now propose to name this other horizon the 'absolute apparent horizon' (AAH). Absolute because even an accelerated ray cannot proceed outward from it.

The reasoning of Hellaby and Krasiński (2002) was as follows. A general null direction $k^\alpha = \mathrm{d}x^\alpha/\mathrm{d}t$ in the metric (2.150) with $\varepsilon = +1$ obeys

$$0 = k^\alpha k^\beta g_{\alpha\beta} = 1 - \frac{(\Phi_{,r} - \Phi\mathcal{E}_{,r}/\mathcal{E})^2}{1-k}\left(\frac{\mathrm{d}r}{\mathrm{d}t}\right)^2 - \frac{\Phi^2}{\mathcal{E}^2}\left[\left(\frac{\mathrm{d}x}{\mathrm{d}t}\right)^2 + \left(\frac{\mathrm{d}y}{\mathrm{d}t}\right)^2\right], \tag{3.114}$$

which implies

$$\frac{(\Phi_{,r} - \Phi\mathcal{E}_{,r}/\mathcal{E})^2}{1-k}\left(\frac{\mathrm{d}r}{\mathrm{d}t}\right)^2 = 1 - \frac{\Phi^2}{\mathcal{E}^2}\left[\left(\frac{\mathrm{d}x}{\mathrm{d}t}\right)^2 + \left(\frac{\mathrm{d}y}{\mathrm{d}t}\right)^2\right]. \tag{3.115}$$

Thus, on a null curve with $dx/dt = 0 = dy/dt$ (which, apart from exceptional cases, will not be a geodesic), the rate of change of r is maximal. One would tend to interpret this to mean that along such a curve the null signal will be able to escape farther than the location of the AH at $\Phi = 2M$. This is in fact not always the case. A curve (3.115) on which $dx/dt = 0 = dy/dt$ will have the maximal dr/dt compared to curves with nonzero values of dx/dt and dy/dt *passing through the point with the same coordinates x and y*. However, going to a different pair of values of x and y changes the value of $\mathcal{E}_{,r}/\mathcal{E}$. For some of the (x, y) pairs this will decrease dr/dt compared to the former (x, y) direction, and the curve (3.115) with $dx/dt = 0 = dy/dt$ will not make it to the AH, being turned back at a smaller value of M. Explicit examples of such a situation will be given elsewhere, here we only wish to give a warning to the readers.

Equation (3.115) implies, along this fastest escape route,

$$\left.\frac{dt}{dr}\right|_n = \frac{j}{\sqrt{1-k}}\left(\Phi_{,r} - \frac{\Phi\mathcal{E}_{,r}}{\mathcal{E}}\right), \qquad j = \pm 1, \tag{3.116}$$

where $j = +1$ for outgoing rays, and $j = -1$ for incoming rays. The solution of the above, $t = t_n(r)$, is the equation of the 'fastest ray'. The value of the function Φ along this ray, $\Phi_n(r) \stackrel{\text{def}}{=} \Phi(t_n(r), r)$, which is an analogue of the Lemaître–Tolman areal radius, will in general be increasing or decreasing with growing r. The absolute apparent horizon (AAH) is where $\Phi_n(r)$ has an extremum, i.e. it changes from increasing to decreasing or vice-versa. Thus, the AAH is a locus where

$$0 = \frac{d\Phi_n}{dr} \equiv \frac{\partial\Phi}{\partial t}\frac{dt_n}{dr} + \frac{\partial\Phi}{\partial r} = \ell j\frac{\sqrt{2M/\Phi - k}}{\sqrt{1-k}}\left(\Phi_{,r} - \frac{\Phi\mathcal{E}_{,r}}{\mathcal{E}}\right) + \Phi_{,r}, \tag{3.117}$$

where $\ell = +1$ for an expanding model and $\ell = -1$ for a collapsing model. Explicitly, this reads[9]

$$\Phi_{,r}\left(\sqrt{1-k} - \sqrt{2M/\Phi - k}\right) + \Phi\sqrt{2M/\Phi - k}\,\mathcal{E}_{,r}/\mathcal{E} = 0. \tag{3.118}$$

Note that the solution of (3.118) is a hypersurface in spacetime determined by the equation $t = t_{AAH}(r, x, y)$. Thus, unlike in the spherically symmetric case or in the case of the ordinary apparent horizon, the function $t_{AAH}(r, x, y)$ essentially depends on three variables and cannot be faithfully represented by a 3D graph.

The equation of the apparent horizon in the (t, M)-coordinates will be the same as in the corresponding L–T model. It is independent of (x, y), and, for the future apparent horizon (in the recollapse phase of evolution of the $k > 0$ model)

---

[9] Note the error in Hellaby and Krasiński (2002): the D in Eq. (169) should in fact be defined by $D = 1 - \sqrt{2M/R + f}/\sqrt{1+f} \equiv 1 - \sqrt{2M/\Phi - k}/\sqrt{1-k}$. Equation (172) is then altogether incorrect, but all the subsequent equations remain unchanged. This error was copied in Sec. 19.7.6 of Plebański and Krasiński (2006), where a similar change should be applied to D, and Eq. (19.243) should be discarded.

it is (see Krasiński and Hellaby, 2004b)

$$t = t_B + \frac{M}{k^{3/2}} \left[\pi + \arccos(-1 + 2k) + 2\sqrt{k(1-k)} \right]. \tag{3.119}$$

In order to find the corresponding equation for the AAH from (3.118) we use the following expression for $\Phi_{,r}$ [to be calculated from the evolution equation, see Sec. 18.10 in Plebański and Krasiński (2006), in particular Eq. (18.107)]:

$$\frac{\Phi_{,r}}{\Phi} = \left(\frac{M_{,r}}{M} - \frac{k_{,r}}{k} \right) + \left(\frac{3}{2} \frac{k_{,r}}{k} - \frac{M_{,r}}{M} \right) \frac{\sin\eta(\eta - \sin\eta)}{(1 - \cos\eta)^2}$$

$$- \frac{k^{3/2}}{M} t_{B,r} \frac{\sin\eta}{(1 - \cos\eta)^2}. \tag{3.120}$$

Before we use this in (3.118) we note that from the evolution equation it follows that with $\pi \leq \eta \leq 2\pi$, where $\sin\eta < 0$, we have

$$\sqrt{2M/\Phi - k} = -\sqrt{k} \frac{\sin\eta}{1 - \cos\eta}. \tag{3.121}$$

We substitute (3.120) and (3.121) in (3.118) and get:

$$\left[\frac{M_{,r}}{M} - \frac{k_{,r}}{k} + \left(\frac{3}{2} \frac{k_{,r}}{k} - \frac{M_{,r}}{M} \right) \frac{\sin\eta(\eta - \sin\eta)}{(1 - \cos\eta)^2} - \frac{k^{3/2}}{M} t_{B,r} \frac{\sin\eta}{(1 - \cos\eta)^2} \right]$$

$$\times \left[\sqrt{1-k} + \frac{\sqrt{k}\sin\eta}{1 - \cos\eta} \right] + \frac{\sqrt{k}\sin\eta}{1 - \cos\eta} \frac{\mathcal{E}_{,r}}{\mathcal{E}} = 0. \tag{3.122}$$

We will use this as an implicit definition of $\eta(M)_{\rm AAH}$, i.e. as the equation of the AAH in the (η, M) variables (η will depend on x and y via $\mathcal{E}$). Then $t(M)$ on the AAH can be found from the evolution equation:

$$t(M)_{AAH} = \left[\frac{M}{k^{3/2}} (\eta - \sin\eta) + t_B \right]_{AAH}. \tag{3.123}$$

Equation (3.122) can be solved only numerically. To verify its solvability, and to find the initial values for the bisection method, we will transform it because in the form (3.122) its left-hand side becomes infinite at $\eta \to 2\pi$.

We will assume that the shell crossings are absent. Among the conditions for no shell crossings, found by Hellaby and Krasiński (2002), the following are useful here:

$$2\pi \left(\frac{3}{2} \frac{k_{,r}}{k} - \frac{M_{,r}}{M} \right) - \frac{k^{3/2}}{M} t_{B,r} < 0 \tag{3.124}$$

(see Eq. (126) in Hellaby and Krasiński, 2002), and

$$\frac{M_{,r}}{M} - \frac{k_{,r}}{k} > 0, \tag{3.125}$$

which follows from the fact that $\Phi_{,r}/\Phi > 0$ must hold for all (η, r), and from (3.120) taken at $\eta = \pi$ (see Hellaby and Krasiński, 2002).

We observe that

$$\lim_{\eta \to 2\pi} \frac{\sin \eta}{\sqrt{1 - \cos \eta}} = -\sqrt{2}, \tag{3.126}$$

and multiply (3.122) by $(1 - \cos \eta)^2$ to obtain:

$$\Psi(\eta) \overset{\text{def}}{=} \left[\left(\frac{M_{,r}}{M} - \frac{k_{,r}}{k} \right) (1 - \cos \eta)^{3/2} \right.$$
$$\left. + \left(\frac{3}{2} \frac{k_{,r}}{k} - \frac{M_{,r}}{M} \right) \frac{\sin \eta (\eta - \sin \eta)}{\sqrt{1 - \cos \eta}} - \frac{k^{3/2}}{M} t_{B,r} \frac{\sin \eta}{\sqrt{1 - \cos \eta}} \right]$$
$$\times \left[\sqrt{1 - k} \sqrt{1 - \cos \eta} + \frac{\sqrt{k} \sin \eta}{\sqrt{1 - \cos \eta}} \right] + \sqrt{k} \sin \eta (1 - \cos \eta) \frac{\mathcal{E}_{,r}}{\mathcal{E}} = 0. \tag{3.127}$$

Now we verify that

$$\lim_{\eta \to \pi} \Psi(\eta) = 4\sqrt{1 - k} \left(\frac{M_{,r}}{M} - \frac{k_{,r}}{k} \right) > 0, \tag{3.128}$$

being positive in consequence of (3.125); and

$$\lim_{\eta \to 2\pi} \Psi(\eta) = 2\sqrt{k} \left[2\pi \left(\frac{3}{2} \frac{k_{,r}}{k} - \frac{M_{,r}}{M} \right) - \frac{k^{3/2}}{M} t_{B,r} \right] < 0, \tag{3.129}$$

being negative in consequence of (3.124).

Thus $\Psi(\pi) > 0$ and $\Psi(2\pi) < 0$, so there exists an $\eta_0 \in (\pi, 2\pi)$ at which $\Psi(\eta_0) = 0$. By this opportunity, we have proved that each particle in a recollapsing quasi-spherical Szekeres model must cross the AAH before it hits the Big Crunch at $\eta = 2\pi$. (Whether η_0 is uniquely defined is not known to us. If it is not, then Ψ may have more than one zero.)

We briefly recapitulate some of the properties of the AAH found in Hellaby and Krasiński (2002) and reported in Plebański and Krasiński (2006); see there for more details (but note: those two references use the name 'apparent horizon' for the entity that we call 'absolute apparent horizon' here).

The absolute apparent horizon may either (1) not intersect a given surface of constant (t, r); or (2) have a single point in common with it; or (3) intersect it along a circle or a straight line. Let

$$D \overset{\text{def}}{=} 1 - \frac{\sqrt{2M/\Phi - k}}{\sqrt{1 - k}}. \tag{3.130}$$

Except for the special case when $S_{,r}/S = D\Phi_{,r}/\Phi$, the circle of intersection in the (x, y) plane has its centre at

$$(x_{AH}, y_{AH}) = \left(P - \frac{P_{,r}}{S_{,r}/S - D\Phi_{,r}/\Phi}, Q - \frac{Q_{,r}}{S_{,r}/S - D\Phi_{,r}/\Phi} \right), \tag{3.131}$$

and the radius $L_{AH} = \sqrt{\lambda}/(S_{,r}/S - D\Phi_{,r}/\Phi)$. The special case $S_{,r}/S = D\Phi_{,r}/\Phi$ [when the locus of the AAH in the (x, y) plane is a straight line] is an artefact of the stereographic projection; this straight line is an image of a circle on the sphere.

From the definition of the absolute apparent horizon, and from $\Phi > 0$, $\mathcal{E} > 0$ and $\Phi_{,r} > 0$ we have

$$(D > 0) \Longrightarrow (\mathcal{E}_{,r} < 0), \qquad (D < 0) \Longrightarrow (\mathcal{E}_{,r} > 0). \tag{3.132}$$

But $D > 0$ and $D < 0$ define regions independent of x and y. Hence, where $D > 0$ (resp. $D < 0$), we have $\mathcal{E}_{,r} < 0$ (resp. $\mathcal{E}_{,r} > 0$) on the whole of the AAH. This implies that the $\mathcal{E}_{,r} = 0$ circle and the AAH cannot intersect unless they coincide. Indeed, these circles lie in parallel planes.

Along $\Phi = 2M$,

$$(\Phi_n)_{,r} = -\Phi \mathcal{E}_{,r}/\mathcal{E}, \tag{3.133}$$

so the apparent horizon, which is at $\Phi = 2M$, does not coincide with the AAH except where $\mathcal{E}_{,r} = 0$.

PART II

Applications of the models in cosmology

4

Structure formation

The formalisms most widely used to describe structure formation and evolution are N-body simulations and the linear perturbation approach. However, in N-body simulations the interactions between particles are described by Newtonian mechanics where, at variance with GR, matter does not affect light propagation, there is instantaneous action at a distance, and there are no curvature effects. As regards the linear approach, we have discussed its drawbacks in Chapter 1.

To avoid such weaknesses, we now apply the methods of Chapter 2, for constructing Lemaître–Tolman, Lemaître and Szekeres models, to several explicit descriptions of structure formation, based on data from actual galaxies, clusters and voids in the observed Universe. Our approach requires certain data to be set at initial and final times, t_1 and t_2, and then calculates the model that evolves between them. Thus, we next consider the observational constraints at recombination, which we will use as an initial time. For constraints at t_2, we will look to present-day observations of relatively nearby structures.

4.1 Initial conditions

4.1.1 Transforming scales in the background

It is a common-sense assumption that present-day cosmic structures evolved from small initial fluctuations whose traces can be observed in the CMB temperature fluctuations. We imagine that a condensed structure at t_2 has accumulated its present mass by drawing matter in from the surroundings. Thus, a condensation (such as a galaxy cluster) will be enveloped in a region where the density is lower than the cosmic average, up to the distance R_c, called the *compensation radius*, at which the total mass within R_c is the same as it would be in a Friedmann (dust) model. Beyond R_c, the Friedmann background remains unperturbed.

For simplicity, in this subsection we assume that the background outside R_c is the $k = 0$ Friedmann model. We wish to know what angle would be subtended in the CMB sky by matter that today forms various objects. For this, we calculate the area or diameter distance (i.e. the areal radius of the past null cone)

$$D_A(t) = S(t) \int_t^{t_2} \frac{1}{S(t')} \, \mathrm{d}t' = 3 \left(t_2^{\,1/3} t^{2/3} - t \right), \tag{4.1}$$

where $S(t) = \left(9M_0 t^2/2\right)^{1/3}$ is the Friedmann scale factor; M_0 is a constant and we assume $t_B = 0$ for the time of the Big Bang. Consequently, an observed area of angular diameter θ on the CMB sky has a physical diameter at the time of recombination t_1 given by

$$L_1 = D_A(t_1)\theta = 3\left(t_2^{1/3} t_1^{2/3} - t_1\right)\theta. \tag{4.2}$$

The present-day diameter of that area – assuming it does not collapse – is thus

$$L_2 = L_1 S_2/S_1. \tag{4.3}$$

We assume that today's structures had not yet existed at $t = t_1$ – their matter was (very nearly) uniformly spread through the background. To calculate the radius occupied by that matter we take the present mass M_2 of the object under consideration and spread it over such a volume that the mean density in the volume becomes equal to the mean background density at t_2. The radius of that region (call it the 'equivalent background radius', EBR) is

$$L_2 = \left(\frac{3M_2}{4\pi\rho_{b2}}\right)^{1/3}. \tag{4.4}$$

We then scale down this radius to t_1 by the law (4.3) and apply (4.2). The result is

$$\left(\frac{3M_2}{4\pi\rho_{b2}}\right)^{1/3}\left(\frac{t_1}{t_2}\right)^{2/3} = 3\left(t_2^{1/3} t_1^{2/3} - t_1\right)\theta. \tag{4.5}$$

This relation is used to calculate the angles given in Table 4.1 (Krasiński and Hellaby, 2002).[1] The symbol in parentheses in the left column of the table defines the unit used to measure the given quantity; ρ_b is the average density of the whole (background) Universe.

Thus, at the current best resolution, the only structures whose trace has a chance of appearing on the CMB map are large cluster concentrations like the Great Attractor. There are just no data to constrain the fluctuations at t_1 that will later develop into smaller structures such as galaxies. Moreover, if on smaller scales the temperature fluctuations are of amplitude $\Delta T/T \leq 10^{-6}$ then the signal will be significantly distorted by the Rees–Sciama (1968) effect (see Sec. 7.1).

4.1.2 Initial density and velocity fluctuations

With the current resolution, the largest amplitude of the temperature fluctuations (except for the dipole) is around 2×10^{-4}. However, if averaged, the amplitude of the rms temperature fluctuations depends on the angular scale. On the scales observed by the WMAP ($\theta > 0.5°$) the rms temperature fluctuations are of amplitude 10^{-5}.

---

[1] Values given here are updated with respect to the 2002 paper using newer data.

Table 4.1. *Basic characteristics of some astronomical objects and their angular diameters on the CMB sky. The parameter* θ_{EdS} *is the angular diameter in the Einstein–de Sitter model [using (4.5)],* $\theta_{\Lambda\mathrm{CDM}}$ *is the angular diameter in the* $\Lambda\mathrm{CDM}$ *model; please note that the angular resolution of the WMAP is approximately* 0.5°*. In the table below, 'angle' means 'angular diameter' (not radius).*

	Object					
	Star	Globular cluster	Galaxy	Virgo cluster	Great Attractor	Void
Radius today (kpc)	$2.3 \cdot 10^{-11}$	0.1	15	2 200	20 000	20 000
Equivalent background rad. (kpc)	0.18	8.35	835	19 000	36 000	14 000
Mass $(M_\odot)$	1	10^5	10^{11}	$1.2 \cdot 10^{15}$	$8 \cdot 10^{15}$	$5 \cdot 10^{14}$
Average dens. (ρ_b)	$1.5 \cdot 10^{29}$	$2 \cdot 10^5$	$6 \cdot 10^4$	190	1.6	0.1
Angle on CMB sky, θ_{EdS} (°)	$1.67 \cdot 10^{-6}$	$7.8 \cdot 10^{-5}$	$7.8 \cdot 10^{-3}$	0.178	0.335	0.133
Angle on CMB sky, $\theta_{\Lambda\mathrm{CDM}}$ (°)	$9.80 \cdot 10^{-7}$	$4.5 \cdot 10^{-5}$	$4.5 \cdot 10^{-3}$	0.100	0.195	0.078

The currently accepted limit on the relative density fluctuation of baryonic matter $\Delta\rho/\rho$ at t_1 is 10^{-5} (Padmanabhan, 1996). To calculate the allowed relative velocity fluctuation we use the result by Dunsby (1997), converted as in Krasiński and Hellaby (2004a) and obtain[2] $\Delta b/b\,(t_1) \approx 10^{-4}$.

No data exist on the *shape* of the density or velocity profiles at t_1 or on the spatial extent of the region that is perturbed at t_1. For the shape, we assume simple functions. It turns out that the final outcome of evolution is sensitive to the initial velocity profile, see Sec. 4.4.

4.2 The galaxy plus black hole formation model

Our aim is to model the formation of a galaxy with a central black hole, starting from an initial fluctuation at recombination. Our model consists of two parts

---

[2] The numbers given here differ from those given by Krasiński and Hellaby (2004a). The differences are caused by an improvement in observational data precision that occurred in the meantime.

joined together across a comoving boundary $M = M_{BH}$, with M_{BH} the estimated present-day mass inside the black hole horizon. In the exterior part, we take existing observational data for the present-day density profile, and the initial fluctuation is made compatible with CMB observations. For the interior, no observational constraints exist, so we propose two possible descriptions, as detailed below. These are both L–T models, and represent a collapsing body, and a dense Kruskal–Szekeres wormhole in the sense of Hellaby (1987).

4.2.1 A galaxy with a central black hole

It has become generally accepted that most large galaxies contain central black holes (e.g. Melia and Falcke, 2001; Begelman, 2003; see Krasiński and Hellaby, 2004b, for a full list). In Secs. 4.2.1–8 we will show how the formation of such a galaxy can be described by an L–T model adapted to this purpose. This is a summary of the research carried out by Krasiński and Hellaby (2004a).

Although spiral galaxies are not spherically symmetric, both the core and the halo – together containing more mass than the disk – are quite close to it, so the L–T model may be an acceptable first approximation. One might consider more general shapes of galaxies using the Szekeres (1975a, 1975b) model, which was successfully applied to studying clusters and voids (Bolejko, 2006a, 2007, and see also Sec. 4.6). This is a realistic task that might (and perhaps will) be undertaken soon. Real galaxies rotate, and to consider the dynamical process of formation of a rotating galactic black hole by methods of exact relativity we would have to use a nonstationary solution with rotating matter. Such solutions, with realistic matter models, are so far unknown (Krasiński, 1997; Stephani *et al.*, 2003; Plebański and Krasiński, 2006), and, considering the difficulties of the task, will probably not be known for a long time to come. So, for the time being, we must be satisfied with the simpler solutions that are available. The real galaxy-plus-black-hole that will provide our data will be M87.

The present state of the galaxy is defined by a mass distribution that consists of two parts:

1. The part outside the apparent horizon at t_2 – for which we use an approximation to the observationally determined density profile of the M87 galaxy. This part extends inward to a sphere of mass M_{BH} – the observationally determined mass of the black hole. Within this section we shall use M_{BH} to mean $M_{BH}(t_2)$, as indicated in Fig. 3.1.

2. The part inside the apparent horizon at t_2. Since, for fundamental reasons, no observational data exist for this region apart from the value of M_{BH}, we are free to choose any geometry. We choose two examples:

2a. A simple subcase of the L–T model, discussed in Sec. 3.2.10 as an illustrative example of properties of horizons. In this model, the black hole does not exist initially and is formed in the course of evolution.

2b. A pre-existing wormhole, also chosen arbitrarily for simplicity of the calculations.

The boundary between the 'inside' and 'outside' at times other than t_2 goes along a comoving mass shell, so that at $t < t_2$ the apparent horizon resides in the inside part.

For the initial time t_1, it is natural to use the last scattering epoch, and this is what we attempted. However, at the angular scales needed for the present calculation, no usable observational data exist. Using the data that are available required some 'creative bookkeeping', which is described at the end of subsection 4.2.5.

These two states, at t_1 and t_2, uniquely define the L–T model that evolves between them, as shown in Sec. 2.1.5. The 3D surface graphs of density as a function of mass and time in Figs. 4.4 and 4.6 show that the evolution proceeds without shell crossings, and so the model is acceptable, at least qualitatively.

4.2.2 The black hole interior

Astronomical observations do not say anything about that portion of galactic matter that had already fallen inside the apparent horizon (AH) by the time the electromagnetic signal that would reach the observer was emitted. What can be seen in the sky are only electromagnetic waves emitted by objects that were still outside the AH at the time of emission. Consequently, we are not constrained in any way in choosing a model for the matter in the interior of a black hole, except for the need to match it smoothly to a galaxy model.

The term 'black hole' is used in two ways. Firstly, there is the Schwarzschild–Kruskal–Szekeres black hole, which has the topology of two universes joined by a temporary wormhole, and begins its life as a white hole.

This was generalised to a matter-filled version in Hellaby (1987), and further generalised in Hellaby and Krasiński (2002). Secondly, there is a black hole formed by the collapse of a massive body, which has an ordinary topology without a wormhole.

Only in their late stages (after the closure of the wormhole) do these two become essentially the same. Both of these can be reproduced by an L–T model, and we consider them in turn below.

At this point, we must make a digression about the relation between model black holes such as those considered here, and real collapsed objects that are called 'black holes' by astronomers. There are two important points to be remembered:

1. The matter proceeding toward a black hole disappears from the field of view of any real observer before it hits the apparent horizon. Therefore, the observationally determined 'mass of a black hole' is in fact only an upper limit of the mass that has actually fallen within the apparent horizon; the latter can never be measured in reality. *Models* of black holes allow us to *calculate* better estimates of that mass, and even if the arithmetic difference between a model calculation and an observational limit is small, it is important to understand the conceptual difference.

2. It is incorrect to speak about the event horizon in the context of observations. We may know where the event horizon is only in a *model*. In practice, we would have to be able to take into account the future fate of every piece of matter, including those pieces that have been outside our field of view up to now – an obviously impossible task. Worse still, if the real Universe is to recollapse in the future, we might already be inside the event horizon and will never see any signature of it. Hence, the horizons whose signatures we have any chance to see (like disappearance of matter from sight, or a large mass being contained in a small volume) are *apparent horizons* – they are local entities, detectable in principle at any instant (although with the difficulty mentioned in point 1). Even in our simple model considered in Sec. 3.2.10 it is rather difficult to determine the position of the event horizon, and it can be done only numerically.

In the following we will identify the estimated black hole mass, obtained by fitting a model to present-day observations, with M_{BH}, the mass within the apparent horizon at time t_2 (see (3.68)).

4.2.3 A collapsed body

For this example, we use the model of Sec. 3.2.10, but with different values of the parameters. The functions have already been chosen so that the origin conditions are satisfied and shell crossings avoided. Since $t_C(M)$ is a fast-increasing function, the future singularity in the black hole forms far earlier than the Big Crunch in the surrounding universe.

To assure a smooth match to the exterior model, we require the continuity of the L–T arbitrary functions and their derivatives at $M = M_{BH}$. Taking the forms of $t_B(M)$ and $E(M)$ given in (3.65) and (3.67), we solve for the constants a, b, T_0 and t_{B0}, so that E, t_B, dE/dM and dt_B/dM are matched at the boundary:

$$t_{B0} = \left[t_B - \frac{M}{2} \frac{dt_B}{dM} \right]_{M=M_{BH}}, \tag{4.6a}$$

$$b = - \left[\frac{1}{2M} \frac{dt_B}{dM} \right]_{M=M_{BH}}, \tag{4.6b}$$

$$T_0 = \left[\frac{M}{6} \frac{dt_B}{dM} + \frac{4\pi M}{3(-2E)^{3/2}} - \frac{2\pi M^2}{(-2E)^{5/2}} \frac{dE}{dM} \right]_{M=M_{BH}}, \tag{4.6c}$$

$$a = \left[\frac{1}{3M^2} \frac{dt_B}{dM} + \frac{2\pi}{3M^2(-2E)^{3/2}} + \frac{2\pi}{M(-2E)^{5/2}} \frac{dE}{dM} \right]_{M=M_{BH}}. \tag{4.6d}$$

These conditions actually ensure more smoothness than is required by the Darmois junction conditions.

4.2.4 A wormhole

Since we have no way of knowing anything about the matter and spacetime interior to M_{BH}, we can equally well fit in a dust-filled wormhole of the Kruskal–Szekeres type, constructed with the L–T metric. The essential requirement is that, at the middle of the wormhole, M must have a minimum value $M_{\min}$, and $E(M_{\min}) = -1/2$. The minimum lifetime (time from past to future singularity) of the wormhole is then $2\pi M_{\min}$. We choose the following functions:

$$t_B = t_{B0} - b(M - M_{\min})^2, \tag{4.7a}$$

$$E = \frac{-M_{\min}}{2M} + a(M - M_{\min}). \tag{4.7b}$$

From these, the conditions for matching to an exterior at some M are

$$M_{\min} = \left[M^2 \frac{\mathrm{d}E}{\mathrm{d}M} + M \left(1 \pm \sqrt{1 + 2E + M^2 \left(\frac{\mathrm{d}E}{\mathrm{d}M} \right)^2} \right) \right]_{M=M_{BH}}, \tag{4.8a}$$

$$a = \left[\frac{1}{2} \frac{\mathrm{d}E}{\mathrm{d}M} - \frac{1}{2M} \left(1 \pm \sqrt{1 + 2E + M^2 \left(\frac{\mathrm{d}E}{\mathrm{d}M} \right)^2} \right) \right]_{M=M_{BH}}, \tag{4.8b}$$

$$b = \left[\frac{\mathrm{d}t_B}{\mathrm{d}M} \bigg/ \left\{ 2M \left(M \frac{\mathrm{d}E}{\mathrm{d}M} \pm \sqrt{1 + 2E + M^2 \left(\frac{\mathrm{d}E}{\mathrm{d}M} \right)^2} \right) \right\} \right]_{M=M_{BH}}, \tag{4.8c}$$

$$t_{B0} = \left[t_B + \frac{M}{2} \frac{\mathrm{d}t_B}{\mathrm{d}M} \left(M \frac{\mathrm{d}E}{\mathrm{d}M} \pm \sqrt{1 + 2E + M^2 \left(\frac{\mathrm{d}E}{\mathrm{d}M} \right)^2} \right) \right]_{M=M_{BH}}. \tag{4.8d}$$

This model was chosen for simplicity, and so is not very flexible. The matching fixes the value of $M_{\min}$, which determines the lifetime of the wormhole. A different model would allow the wormhole lifetime to be a free parameter. The apparent horizons and singularities for this model with the parameters (4.20) used below are illustrated in Fig. 4.1.

4.2.5 The exterior galaxy model

The final density profile

As our example for the density profile of the final state, we choose the galaxy M87. It is believed to contain a large black hole (Ford *et al.*, 1994; Harms *et al.*, 1994; Macchetto *et al.*, 1997) and the density profile for its outer part had been

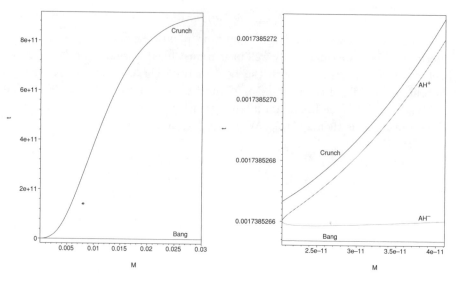

Fig. 4.1. The L–T model for the wormhole interior of Sec. 4.2.4 with parameters (4.8), their values in geometric units given by (4.20). The second graph
shows the neck of the wormhole magnified about 10^9 times. Only on this scale
are the apparent horizons distinguishable from the bang and crunch. The bang
does actually curve downwards, but its variation is much less than that of the
crunch. The neck is at the left where the mass reaches a minimum $M_{\min}$. The
apparent horizons cross in the neck at the moment of maximum expansion.
The second sheet is not shown. It could be a mirror image of this sheet,
or it could be quite different. The boundary with the exterior model (at
$M_{BH} = 0.03$) is defined to be where $t_2 = t_{AH}$. Recombination t_1 is indistinguishable from the bang in the first graph, and far to the future in the second.

proposed some time ago (Fabricant *et al.*, 1980):

$$\rho(s) = \rho_0 / \left(1 + bs^2 + cs^4 + ds^6\right)^n, \qquad (4.9)$$

where $\rho_0 = 1.0 \cdot 10^{-25}$ g/cm^3, $b = 0.9724$, $c = 3.810 \cdot 10^{-3}$, $d = 2.753 \cdot 10^{-8}$, $n = 0.59$. The distance from the centre, s, is measured in arc minutes, i.e. it is dimensionless, and is related to the actual distance r by

$$s = \frac{r}{D} \frac{10800}{\pi} \overset{\text{def}}{=} r\delta, \qquad (4.10)$$

where D is the distance from the Sun to the galaxy. (Of course, no galaxy and no
real black hole is spherically symmetric, so we cannot model any actual galaxy
with the L–T solution. However, we wish to make our illustrative example as
close to reality as possible, and this is why we stick to an actual object.) For our
purposes, we need density expressed as a function of mass, and a profile that
goes to infinity at $r \to 0$, to allow for the singularity inside the black hole.[3] The

---

[3] Even though we will use the interior model near the centre, this requirement assists in joining
interior and exterior smoothly.

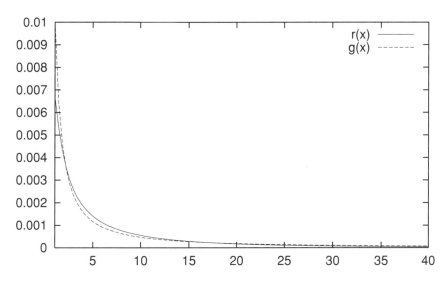

Fig. 4.2. Comparison of the profile (4.9) (marked as $r(x)$) with (4.11) (marked as $g(x)$). The ρ on the vertical axis is mass density in units of 10^{-23} g/cm^3, the x on the horizontal axis is the distance from the centre in arc minutes, just as in Fabricant *et al.* (1980). The values of parameters are given below (4.9). Note: the lowest value of $s \equiv \delta r$ on the horizontal axis is 1.

mass profile corresponding to (4.9) is not an elementary function. However, it turns out that the following very simple profile is a close approximation to (4.10) in the region considered by Fabricant *et al.* (1980) (see Fig. 4.2):

$$\rho(r) = \rho_0/(\delta r)^{4/3}, \tag{4.11}$$

with the same value of ρ_0.

The corresponding mass distribution is

$$\widetilde{M}(r) = \frac{12}{5}\pi\rho_0 r^{5/3}\delta^{-4/3}, \tag{4.12}$$

and we make it singular at $r = 0$ by adding a constant M_S to $\widetilde{M}(r)$ (this is the same M_S as defined before – the mass that had already fallen into the singularity at t_2; so far this is still an arbitrary constant):

$$M(r) = M_S + \frac{12}{5}\pi\rho_0 r^{5/3}\delta^{-4/3}. \tag{4.13}$$

Hence, $r = [5(M - M_S)\delta^{4/3}/12\pi\rho_0]^{3/5}$, and[4]

$$\rho(M) = (12\pi/5)^{4/5}\rho_0^{9/5}\delta^{-12/5}(M - M_S)^{-4/5}. \tag{4.14}$$

---

[4] Note that adding M_S in (4.13) does not change the function $\rho(r)$ in (4.11), although it changes $\rho(M)$.

In principle, the value of ρ near to $M = M_{BH}$ should be measurable, but in this regime the difference between the Newtonian and L–T definitions of density becomes too pronounced (because of the non-flat geometry). To infer $\rho(M_{BH})$ in a sensible way, the results of observations should be consistently reinterpreted within the L–T scheme, and such results are not, and will not be available for a long time. Consequently, we have to give up on this bit of information.

From (2.17), we find the corresponding $R(M)$ via $R^3 = (3/4\pi) \int_{M_S}^{M} dx/\rho(x)$, which is

$$R(M) = \left(\frac{5}{12\pi\rho_0} \right)^{3/5} \delta^{4/5} (M - M_S)^{3/5} . \tag{4.15}$$

Now we can determine M_S by the requirement that $R = 2M$ at $M = M_{BH}$, i.e.

$$\left(\frac{5}{12\pi\rho_0} \right)^{3/5} \delta^{4/5} (M_{BH} - M_S)^{3/5} = 2M_{BH}. \tag{4.16}$$

From here we find

$$M_S = M_{BH} - \frac{12\pi\rho_0}{5\delta^{4/3}} (2M_{BH})^{5/3}. \tag{4.17}$$

It follows, as it should, that $M_S < M_{BH}$. However, this result makes sense only if $M_S > 0$, which is equivalent to

$$\left(\frac{24}{5}\pi\rho_0 \right)^{3/2} \cdot \frac{2M_{BH}}{\delta^2} < 1. \tag{4.18}$$

For checking this inequality, all quantities have to be expressed in geometric units. The black hole in M87 is believed to have mass $M_{BH} = 3 \cdot 10^9 M_\odot$ (Di Matteo *et al.*, 2003; Macchetto *et al.*, 1997), and its distance from the Sun is $D = 17$ Mpc. In geometric units, with 1 year $= 31\,557\,600$ s, $c = 3 \cdot 10^9$ cm/s, $G = 6.6726 \cdot 10^{-8}$ cm^3/g$\cdot$s^2 and $M_\odot = 1.989 \cdot 10^{33}$ g, this makes $M_{BH} = 4.424 \cdot 10^{14}$ cm, $\delta = 6.553 \cdot 10^{-23}$ cm^{-1}, $\rho_0 = 0.741 \cdot 10^{-53}$ cm^{-2}, and the left-hand side of (4.18) comes out to be $2.437 \cdot 10^{-19}$, which is very safely within the limit.

The initial fluctuation

In order to define a model uniquely, we only need one more profile for the region $M \geq M_{BH}$, e.g. a density or velocity profile at $t = t_1 < t_2$ or a specific choice of $E(M)$ or $t_B(M)$. Apart from the density profile at $t = t_2$, there are no other observational constraints in the region $M \geq M_{BH}$ – the time by which galaxies started forming is not well known, and presumably different for each galaxy, nothing is known about the initial density or velocity distribution in the proto-galaxy at that time.

Since the only quantities that are to some degree constrained by the observations are the density and velocity profiles at the recombination epoch, it is most natural to use these for t_1. Even so, there is a problem: no numerical data are available for amplitudes of the temperature fluctuations of the CMB radiation at such small scales. The expected angular size on the CMB sky of a perturbation that will develop into a single galaxy ($0.004°$) is much smaller than the current best resolution ($0.1°$ – see Sec. 4.1.1). Therefore we tried an exactly homogeneous initial density and, in a second numerical experiment, a homogeneous initial velocity. The former turned out to lead to an unacceptable configuration at t_2: a collapsing hyperbolic model with no Big Bang in the past. Consequently, we settled on the homogeneous initial velocity, which then implied the amplitude of the order 10^{-3} for the initial density perturbation.

4.2.6 Numerical evolution of the models

The programs written for the papers by Krasiński and Hellaby (2002 and 2004a) are adapted to facilitate this two-step model construction. First the exterior profiles are used as input to the methods of Sec. 2.1.5, solving numerically for $E(M)$ and $t_B(M)$ for $M_{BH} \leq M \leq M_{\text{galaxy}}$. The values of E and t_B and their derivatives at M_{BH} are extracted, and the parameters of the interior model calculated from them. Then the functions E and t_B are numerically extended into the interior model, down to $M = 0$ or $M = M_{\text{min}}$. From these data, the model evolution is reconstructed using existing programs.

Our first model uses the final density profile of Sec. 4.2.5 for the galaxy at time $t_2 = 14$ Gy, and a flat initial velocity profile at time $t_1 = 10^5$ y, both exterior to M_{BH}. The interior of M_{BH} is a black hole formed by collapse, as described by (3.65) and (3.67), with parameters determined by the matching (4.6). Geometric units are chosen such that $10^{11} M_\odot$ is the mass unit. In these units, the parameters are:

$$a = 2.5 \times 10^{14}, \qquad b = 41525.5859, \qquad t_{B0} = 8043.214,$$
$$T_0 = 8.901 \times 10^{11}. \tag{4.19}$$

The resulting arbitrary functions and the behaviour of the combined model are shown in Figs. 4.3 and 4.4. Notice that the fluctuations of the density at recombination are of the order 10^{-3} (the fluctuations of velocity at recombination are zero by assumption). The black hole singularity forms at time $T_0 = 13.894$ Gy (since $t_{B0} = 125.55$ y is negligible), so it is 106 million years old by today.

Our second model uses the identical exterior, but the interior is a full Kruskal–Szekeres-type black hole containing a temporary 'wormhole', as described by (4.7), with parameters determined by the matching (4.8). The same geometric units are used, and, using the '−' sign in (4.8), the parameters are:

$$a = -4.8597 \times 10^{-8}, \qquad b = 41525.5859, \qquad t_{B0} = 8043.214,$$
$$M_{\text{min}} = 1.8571 \times 10^{-11} \tag{4.20}$$

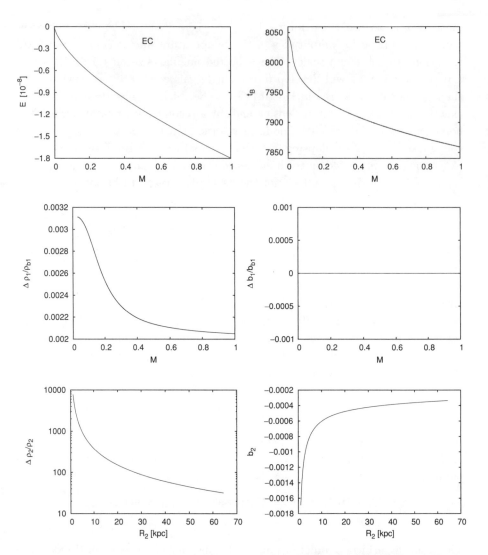

Fig. 4.3. The L–T model for the formation of a galaxy that develops a central black hole. Shown are the model-defining L–T functions $E(M)$ and $t_B(M)$, the $\rho_1(M)$ and $b_1(M)$ fluctuations, the $\rho_2(R_2)$ and $b_2(R_2)$ variations. The graphs in the four lower panels do not show the inside of the apparent horizon. The 'EC' indicates the range considered is an elliptic region that is collapsing by t_2.

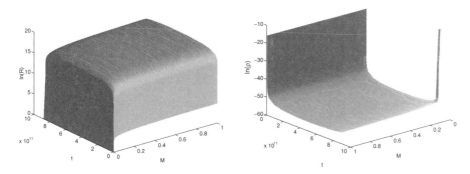

Fig. 4.4. The evolution of the L–T model for the formation of a galaxy that develops a central black hole. Shown are the evolution of $R(t, M)$ and of $\rho(t, M)$. In the $R(t, M)$ graph, the origin $R(t, 0) = 0$ is on the lower left axis, and expansion to a maximum occurs as time increases towards the 'northwest'. At the initial time, $t_1 = $ recombination, $R > 0$ except at the origin. In the $\rho(t, M)$ graph the view has been rotated by $180°$ relative to the $R(t, M)$ graph for clarity, so that the origin is on the right ('behind' the surface graph) and time increases towards the 'south-east'. By $t_2 = $ today, the innermost region has collapsed to a black hole of mass $3 \times 10^9 \, M_\odot$.

(using the '+' sign in (4.8) gives $M_{\min} = 0.06 > M_{BH} = 0.03$, which is not acceptable). Figures 4.5 and 4.6 show the arbitrary functions and the behaviour of the combined model for this scenario.

Because the exteriors are identical, the fluctuations of density at recombination (t_1) outside M_{BH} are again well within CMB limits. The wormhole mass (minimum in M) is $M_{\min} = 1.8571 \, M_\odot$, and the future singularity first forms at $T_0 = 5.7476 \times 10^{-5}$ s after the past singularity. (The future and past black hole singularities are the extension of the crunch and bang into the middle of the wormhole.) The very short lifetime of the wormhole is a consequence of the need for E to go from $-1/2$ all the way up to -1.3435×10^{-9} and arrive there with a negative gradient (see Fig. 4.5, upper left panel). By (2.8), a comparison of the lifetimes at the neck (where $T = T_0$, $M = M_{\min}$ and $E = -1/2$) and the boundary (where $M = M_{BH}$), gives

$$\frac{T_0}{T(M_{BH})} = \frac{M_{\min}}{M_{BH}} \times (2 \times 1.3435 \times 10^{-9})^{3/2} < 8.6 \times 10^{-23},$$

since $M_{\min} < M_{BH}$, which shows the wormhole lifetime must be an extremely small fraction of the galaxy lifetime.

Though models could no doubt be found with quite different wormhole lifetimes, this example very effectively highlights the fact that the two kinds of black holes are observationally indistinguishable, and that the age of the central black hole may be impossible to determine observationally. By recombination (t_1),

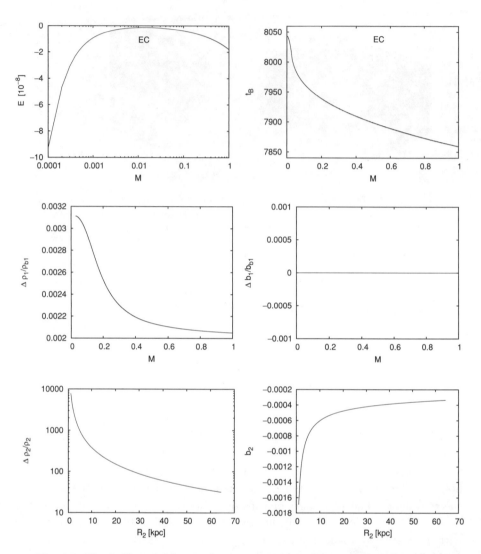

Fig. 4.5. The L–T model for the formation of a galaxy around a pre-existing central black hole. Shown are the model-defining L–T functions $E(M)$ and $t_B(M)$, the $\rho_1(M)$ and $b_1(M)$ fluctuations, the $\rho_2(R_2)$ and $b_2(R_2)$ variations. The graphs in the four lower panels do not show the inside of the apparent horizon.

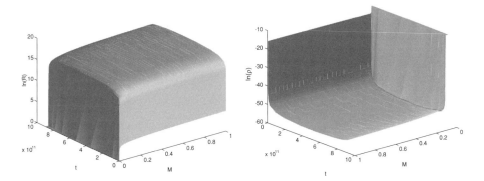

Fig. 4.6. The evolution of the L–T model for the formation of a galaxy around a pre-existing central black hole. Shown are the evolution of $R(t, M)$ and of $\rho(t, M)$. In the $R(t, M)$ graph, on the left there is no origin, rather a wormhole with a tiny lifetime recollapses to zero size. The thin flat wedge along the left is the growing singularity. (The apparent ripples are small numerical variations overemphasised by the graphics program.) In the $\rho(t, M)$ graph, again with the view rotated by 180° relative to the $R(t, M)$ graph, the density diverges towards the singularity, and the flat wedge represents the part that has already collapsed. By $t_2 =$ today, the black hole mass has increased to $3 \times 10^9 \ M_\odot$.

this black hole has accreted[5] 815 091 7 $M_\odot$ within the apparent horizon, which is only 0.161 AU across. Any effect this might have on the CMB will not be observable for a long time.

4.3 Modelling a rich galaxy cluster

4.3.1 The most realistic model

In this model, we choose units in which $c = 1$ and $G = 1$, and we choose the mass unit $1 M_G = 10^{15} M_\odot$. The associated gravitational units of distance and time are $1 L_G = 47.84$ pc and $1 T_G = 156$ y.

One of the objects we choose to model is the galaxy cluster A2199 from the Abell catalogue. The following (Newtonian!) 'universal profile' is in use in astronomy (Navarro *et al.*, 1995):

$$\rho(R) = \rho_b \frac{\delta}{(R/R_s)(1 + R/R_s)^2}, \tag{4.21}$$

where ρ_b ($= 8 \times 10^{-30}$ g/cm^3) is the average density in the Universe, δ ($= 77\,440$) is a dimensionless factor and R_s ($= 3457 L_G = 1.65 \times 10^5$ pc $= 5.11 \times 10^{18}$ km) is a scale distance. This profile is said to apply for R changing by two orders of magnitude (Navarro *et al.*, 1995).

---

[5] The mass within the AH at t_1 is found by numerical root finding, using (3.68) with t_1 instead of t_2 and (3.60).

For our procedure, we need the density given as a function of mass.[6] The calculation $\rho(R) \to M(R) \to R(M) \to \rho(M)$ can always be done numerically, but it is more instructive to have exact explicit formulae. We therefore approximate the 'universal profile' by the following $\rho(M)$ profile [where now ρ is meant to be that of (2.3); indices refer to our various numerical experiments]:

$$\rho_{17}(M) = \rho_b \frac{B_2}{1 + e^{\sqrt{M}/\mu_2}}, \qquad (4.22)$$

in which $B_2 = 498\,500$ and $\mu_2 = 0.07144 M_G$. Figure 4.7 shows the comparison of the profiles (4.21) and (4.22). For Fig. 4.7 the profile (4.22) was numerically recalculated into $\rho(R)$. For information on how the profile (4.22) was guessed see Krasiński and Hellaby (2004a).

We assume that the initial density perturbation contained a mass of about 0.01 of the final mass, so for the initial profile at t_1 we choose

$$\rho/\rho_b = \begin{cases} 1 + A_1 \left[1 + \cos\left(\frac{100\pi M}{M_c} \right) \right] & \text{for} \quad M < M_c/100 \\ 1 & \text{for} \quad M > M_c/100, \end{cases} \qquad (4.23)$$

where $A_1 = 0.000\,015$ and $M_c = 1.182 M_G$ is the mass within the compensation radius, see Fig. 4.7.

Figure 4.8, left graph, shows the numerically calculated function $E(M)$ corresponding to the profile shown in Fig. 4.7. It is everywhere negative, which means that all matter within the model will eventually collapse. The dotted vertical line separates matter that is already recollapsing at t_2 (left) from matter that is still expanding at t_2. Figure 4.8, right graph, shows $t_B(M)$ for the same evolution. The difference in age between the central part of the cluster and its outer part is approx. $0.1 T_G \approx 15.6$ years, i.e. this Big Bang is almost simultaneous. However, strictly speaking, the increasing function $t_B(M)$ implies a shell crossing, but it would occur before t_1, where the model does not apply anyway.

Given $E(M)$ and $t_B(M)$, we can calculate the associated distribution of velocity $b_1(M) = R_{,t}(t_1, M)$ at t_1. This is shown in Fig. 4.9. The relative amplitude of the initial velocity at the centre is about 5×10^{-4}, which is within observational limits (but we recall that these limits correspond to objects of much larger size).

4.3.2 Results of some other numerical experiments

For the evolution from a homogeneous initial density to a galaxy cluster [another profile approximating the 'Universal Profile' (Krasiński and Hellaby, 2004a)] we

---

[6] Note that the relation between mass and density in the L–T model, (2.3), is different from the flat-space relation $dM/dR = 4\pi\rho R^2$. We carry over (4.21) to relativity with the hope that it is a good approximation at low densities. However, a completely self-consistent approach would require re-interpretation of all the relevant astronomical observations against the L–T model. The present work is a step towards that goal.

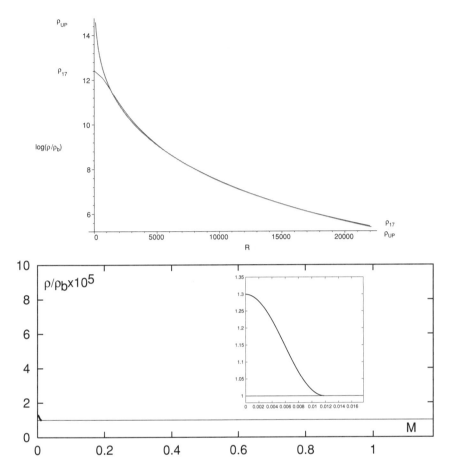

Fig. 4.7. Comparison of the present-day L–T density profile ρ_{17} used for the Abell cluster model with the Newtonian Universal Profile (top panel) and the initial density profile used for the Abell cluster model (bottom panel). The inset shows the enlarged view of the perturbation region close to the centre.

obtained the velocity amplitude at t_1,

$$(\Delta b/b)(t_1)|_{\max} = 5.2 \times 10^{-4}, \tag{4.24}$$

which is on the border of the observationally implied range. This means that a pure velocity perturbation can very nearly produce a galaxy cluster.

For the evolution from a homogeneous initial velocity to a galaxy cluster (same final profile as above) we obtained the density amplitude at t_1,

$$(\Delta \rho/\rho)(t_1)|_{\max} = 12 \times 10^{-3}, \tag{4.25}$$

which differs from the observationally allowed value by 3 orders of magnitude. Hence, velocity perturbations generate structures much more efficiently than density perturbations.

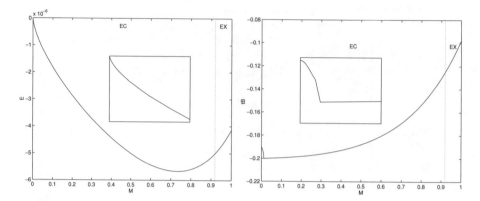

Fig. 4.8. The function $E(M)$ (left) and the bang function (right) for the Abell cluster model. The insets are magnifications of the region near the origin. Explanation in the text.

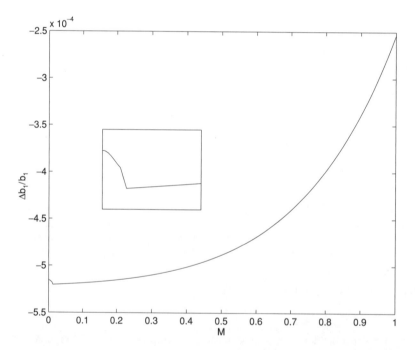

Fig. 4.9. The implied initial velocity distribution for the Abell cluster model. The inset is the magnification of the region near the origin. Explanation in the text.

We demonstrated that a smooth evolution can take an initial condensation to a void (see also Mustapha and Hellaby, 2001). Thus the initial density distribution does not determine the final structure that will emerge from it; the velocity distribution can obliterate the initial setup. This remark should be taken very seriously by the proponents of the classical structure formation paradigm, in which it is assumed that density fluctuations alone are responsible for the creation of structures.

4.4 Modelling a void

In this section we check whether it is possible to evolve a void from small initial density and velocity perturbations imposed on a homogeneous background.

Voids are vast regions where only few instead of thousands of galaxies are observed. According to the astronomical observations, about 40% of the volume of the Universe is taken up by voids (Hoyle and Vogeley, 2004). This means that void formation is not an isolated event, but a very probable process. While dealing with gravity, it is relatively easy to reproduce high-density regions, for instance by setting the initial conditions so that collapse or shell crossings occur. This cannot be done in the case of low-density regions. Thus the study of void formation gives us a better understanding of the Universe in its early stages of evolution.

All models in this section are specified by the initial density and velocity distributions.

4.4.1 The algorithm

The computer algorithm used to calculate the evolution of a void is written in Fortran and consists of the following steps. Numerical methods are from Press *et al.* (1986) and Pang (1997).

1. The initial time t_1 is chosen to be the time of last scattering, and is calculated from the following formula (Peebles, 1980) for a background FLRW universe:

$$t(z) = \frac{1}{H_0} \int_z^\infty \frac{\mathrm{d}\widetilde{z}}{(1+\widetilde{z})\sqrt{\mathcal{D}(\widetilde{z})}}, \qquad (4.26)$$

where:

$$\mathcal{D}(z) = \Omega_\gamma (1+z)^4 + \Omega_m (1+z)^3 + \Omega_K (1+z)^2 + \Omega_\Lambda, \qquad (4.27)$$

H_0 is the present Hubble constant, $\Omega_K = 1 - \Omega_\gamma - \Omega_m - \Omega_\Lambda$, $\Omega_\gamma = (8\pi G a T_{\mathrm{CMB}}{}^4)/(3H_0^2)$, $\Omega_\Lambda = (1/3)(c^2\Lambda/H_0^2)$, $a = 4\sigma/c$, σ is the Stefan–Boltzmann constant, and T_{CMB} is the current temperature of the CMB, 2.75 K. For the lower limit of integration, we used $z = 1089$ (Bennett *et al.*, 2003) as the redshift at last scattering.

2. The initial density and velocity fluctuations, imposed on this homogeneous background, are defined by functions $\delta(\ell)$ and $\nu(\ell)$ of radius ℓ, as listed in

Table 4.2. *The initial density perturbations used in the void formation runs. All the values in the table are dimensionless, and the distance parameter is the areal radius in kiloparsecs, $\ell = R_1/1$ kpc. Note that the output figures depend on the initial perturbations in both density and velocity.*

Model	Density perturbations	Parameters
1	$\delta_1(\ell) = A(-\arctan a + b\ell + ce^{-g^2} + de^{-h^2} + fe^{-j^2}) \cdot k - 1$	$A = -4.4 \times 10^{-5}$, $a = 0.16\ell - 2.2$, $b = 0.036$, $c = 0.125$, $d = 0.25$, $f = 0.35$, $g = \frac{\ell-7}{6}$, $h = \frac{\ell-9}{7}$, $j = \frac{\ell-11}{3}$, $k = \frac{1}{1+0.03\ell}$
2	$\delta_2 = \delta_1(\ell)$	
3	$\delta_3 = 0$	
4	$\delta_4(\ell) = A(-\arctan a + b\ell + ce^{-g^2} + de^{-h^2} - de^{-j^2}) \cdot k - 1$	$A = -3 \times 10^{-3}$, $a = 0.08\ell - 1.1$, $b = 0.023$, $c = 0.1$, $d = 0.25$, $g = \ell/4$, $h = \frac{\ell-2}{7}$, $j = \frac{\ell-4}{3}$, $k = \frac{1}{1+0.03\ell}$
5–7	$\delta_{5-7}(\ell) = 100 \times \delta_1(\ell)$	
8	$\delta_8(\ell) = 2 \times \delta_{5-7}(\ell)$	

Tables 4.2, 4.3[7], and the actual density and velocity followed from

$$\rho_1(\ell) = (\rho_b)_1(1 + \delta(\ell)) \qquad \text{and} \qquad u_1(\ell) = (u_b)_1(1 + \nu(\ell)).$$

The parameter ℓ is defined as the areal radius at the moment of last scattering, measured in kiloparsecs, and is also used for the radial coordinate, i.e.

$$r = \ell = R_1/d = R(r, t_1)/d,$$

where $d = 1$ kpc.

Background values are calculated as follows:

$$(\rho_b)_1 = \rho_0(1 + z_1)^3, \qquad (u_b)_1 = \ell H_0 \sqrt{\mathcal{D}(z_1)},$$

where ρ_0 and H_0 are the current values of the average matter density in the Universe and of the Hubble constant, respectively.

---

[7] The form of these initial density and velocity distributions is a result of consecutive adjustments made in order to test the influence on void formation of the various factors mentioned at the beginning of this section.

Table 4.3. *The initial velocity perturbations used in the void formation runs.
All the values in the table are dimensionless, and the distance parameter is
the areal radius in kiloparsecs,* $\ell = R_1/1$ *kpc.*

Model	Velocity perturbations	Parameters
1	$\nu_1(\ell) = A(-\arctan a + b\ell + ce^{-g^2} - de^{-h^2} - fe^{-j^2}) \cdot k - 1$	$A = 1.6 \times 10^{-4}$, $a = 0.16\ell - 2.2$, $b = 0.036, \ c = 0.125$, $d = 0.25, \ f = 0.35$, $g = \frac{\ell-7}{6}, \ h = \frac{\ell-9}{7}$, $j = \frac{\ell-11}{3}, \ k = \frac{1}{1+0.03\ell}$
2	$\nu_2 = 0$	
3	$\nu_3 = \nu_1(\ell)$	
4	$\nu_4(\ell) = A(-\arctan a + b\ell + ce^{-g^2} - de^{-h^2} - de^{-j^2}) \cdot k - 1$	$A = 3 \times 10^{-3}$, $a = 0.08\ell - 1.1, \ b = 0.023$, $c = 0.1, \ d = 0.25$, $g = \frac{\ell}{4}, \ h = \frac{\ell-2}{7}$, $j = \frac{\ell-4}{3}, \ k = \frac{1}{1+0.03\ell}$
5	$\nu_5(\ell) = 37.5 \cdot \nu_1(\ell)$	
6, 8	$\nu_{6,8}(\ell) = A(-\arctan a + b\ell + ce^{-g^2} + de^{-h^2} + de^{-j^2} + fe^{-n^2}) \cdot k + 5 \times 10^{-4} - 1$	$A = 1.4 \times 10^{-2}$, $a = 0.02\ell - 0.02, \ b = 0.023$, $c = 0.175, \ d = 0.25$, $f = 0.306, \ g = \ell, \ h = \frac{\ell-1}{7}$, $j = \frac{\ell-3}{3}, n = \frac{\ell-39}{12}, \ k = \frac{1}{1+0.03\ell}$
7	$\nu_7(\ell) = A(-\arctan a + b\ell + ce^{-g^2} + de^{-h^2} + de^{-j^2} + fe^{-n^2}) \cdot k + 5 \times 10^{-4} - 1$	$A = 1.4 \times 10^{-2}$, $a = 0.02\ell - 0.02, \ b = 0.023$, $c = 0.175, \ d = 0.25$, $f = 0.175, \ g = \ell, \ h = \frac{\ell-1}{7}$, $j = \frac{\ell-3}{3}, n = \frac{\ell-39}{12}, \ k = \frac{1}{1+0.03\ell}$

Because of lack of precise observational data, it is not possible to model the
profile of the initial density and velocity perturbations. Initial profiles are
assumed to have a minimum of density and a maximum of velocity at the
centre.[8] The chosen initial density and velocity distributions are shown in
Fig. 4.10.

---

[8] However, the results of the previous section and of the works by Mustapha and Hellaby
(2001) and Krasiński and Hellaby (2002) show that maxima and minima can be reversed
during evolution, and it is not at all necessary that a void begins with a minimum of density
at the centre.

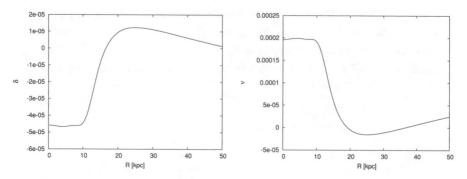

Fig. 4.10. The initial density (left) and velocity (right) perturbations for void-formation model 1 (see Tables 4.2 and 4.3 for details).

3. Then the mass inside a shell of radius R_i, measured in kiloparsecs, is calculated by integrating (2.3):

$$M(\ell) - M(0) = \left. \frac{\kappa c^2}{2} \int_{\ell_{\min}}^{\ell} \rho_i(\ell')\ell'^2 \mathrm{d}\ell' \right|_{t=t_1} . \qquad (4.28)$$

Since the density distribution has no singularities or zeros over extended regions, it is assumed that $\ell_{\min} = 0$ and $M = 0$ at $\ell = 0$.

4. The function E is calculated from R_1, $V = (R_{,t})_1$, M and a chosen Λ value, using (2.2).

5. Then t_B is calculated from (2.4).

6. Once M, E and t_B are known, the state of the L–T model can be calculated for any instant. Solving (2.2) with the second-order Runge–Kutta method for $R(t, \ell)$ along each constant ℓ worldline, we calculate the value of $R(t, \ell)$ and $R_{,t}(t, \ell)$ up to the present epoch.

7. The density $\rho(t, \ell)$ is then found from (2.3) using the five-points differentiating formula. The adjusted differences between adjacent worldlines, used in estimating derivatives, are 10 pc.

8. For the purpose of comparing our results with the observational data the results shown in Figs. 4.11, 4.12, 4.14 and 4.15 do not present the real density contrast, but the average one, i.e.,

$$\delta = \frac{<\rho>}{\bar{\rho}} - 1, \qquad (4.29)$$

where $\bar{\rho}$ is the present background density, and $<\rho> = (3Mc^2)/(4\pi GR^3)$.

9. The current expansion rate inside voids is represented by the equivalent of the Hubble parameter, which is defined as $1/3$ of the expansion scalar (Ellis, 1971):

$$H = \frac{1}{3}\theta = 2\frac{R_{,t}}{R} + \frac{R_{,tr}}{R_{,r}}. \qquad (4.30)$$

Table 4.4. *Summary of results for void formation runs. See Tables 4.2, 4.3 and Figs. 4.10–4.17 for details.*

Model	Ampl. of initial pert.	Description
1	$\delta_i = -4.5 \times 10^{-5}$ $\nu_i = 2 \times 10^{-4}$	Does not reconstruct the present-day voids. Present value: $\delta_0 \approx -0.2$.
2	$\delta_i = -4.5 \times 10^{-5}$ $\nu_i = 0$	Does not reconstruct the present-day voids. Present value: $\delta_0 \approx -0.005$.
3	$\delta_i = 0$ $\nu = 2 \times 10^{-4}$	Results similar to model 1.
4	$\delta_i = -3.7 \times 10^{-3}$ $\nu_i = 3.65 \times 10^{-3}$	Leads to formation of underdense region. Present value: $\delta_0 \approx -0.8$.
5	$\delta_i = -4.5 \times 10^{-3}$ $\nu_i = 7.5 \times 10^{-3}$	Reconstructs the present-day voids. Present value: $\delta_0 \approx -0.94$.
6	$\delta_i = -4.5 \times 10^{-3}$ $\nu_i = 8 \times 10^{-3}$	Reconstructs the present-day voids. Present value: $\delta_0 \approx -0.94$.
7	$\delta_i = -4.5 \times 10^{-3}$ $\nu_i = 8 \times 10^{-3}$	Reconstructs the present-day voids. Present value: $\delta_0 \approx -0.94$.
8	$\delta_i = -9 \times 10^{-3}$ $\nu_i = 8 \times 10^{-3}$	Reconstructs the present-day voids. Present value: $\delta_0 \approx -0.94$.

4.4.2 Void formation

The evolution of models 1–4 is presented in Fig. 4.11. Table 4.4 presents a short description of the results of these models. As can be seen, density fluctuations of amplitude $\sim 10^{-5}$ and velocity fluctuations of amplitude $\sim 10^{-4}$ are insufficient to reproduce the currently observed low-density regions (models 1–3). The necessary amplitude is at least $\sim 10^{-3}$ (model 4). Another interesting result is that density fluctuations alone are not responsible for void formation – in model 2 where $\nu = 0$ the current density contrast is $\delta_0 \approx -0.005$. Thus, as follows from the comparison of model 3 and model 1, velocity fluctuations are of greater importance.

Analysis of the evolution of cosmic voids in different cosmological models implies similar results. Figure 4.12 presents void evolution in six different background models – initial conditions as in model 4. Some of these models, especially the elliptic ones with and without the cosmological constant, are inconsistent with observations. These are not intended as realistic models of observed voids, but rather as confirmation that the above results are not background specific. Curves e and f in Fig. 4.12 are truncated. At the cutoff point a shell crossing

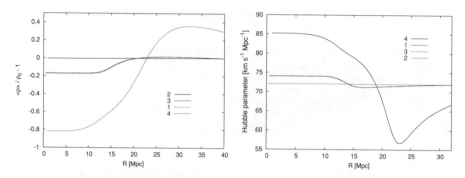

Fig. 4.11. The current average density contrast (left) and the Hubble parameter (right) for void-formation models 1–4.

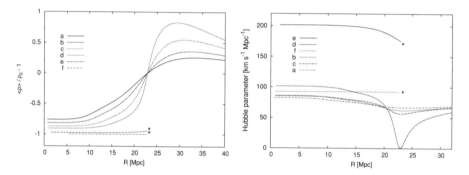

Fig. 4.12. The current density contrast and the Hubble parameter for void-formation model 4 in six different backgrounds models. (a) $\Omega_m = 0.27$, $\Omega_\Lambda = 0$; (b) $\Omega_m = 0.39$, $\Omega_\Lambda = 0$; (c) $\Omega_m = 0.27$, $\Omega_\Lambda = 0.73$; (d) $\Omega_m = 1$, $\Omega_\Lambda = 0$; (e) $\Omega_m = 11$, $\Omega_\Lambda = 0$; (f) $\Omega_m = 0.27$, $\Omega_\Lambda = 1.64$. * – shell crossing.

occurs – the inner shells catch up with the outer shells. This results in a singularity that probably does not occur in the real Universe. Before it happens, the gradient of pressure would become significant and the L–T model would become inapplicable, see Bolejko and Lasky (2008).

As can be seen from Fig 4.12, the major influence of the background on structure formation is from the shell expansion. Except for the model of curve f ($\Omega_m = 0.27$ and $\Omega_\Lambda = 1.64$), where the age of the Universe is almost 32×10^9 y, the higher the expansion rate the deeper the void is. The fastest expansion rate is in the model of curve e ($\Omega_m = 11$, $\Omega_\Lambda = 0$; age of the Universe $\approx 4.6 \times 10^9$y), which leads to the most negative present-day density contrast.

4.4.3 Void evolution

Let us now focus on realistic models of void evolution: models 5–8. The initial fluctuations used in this subsection are presented in Fig. 4.13, the profile

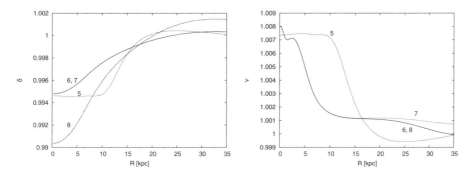

Fig. 4.13. The initial density (left) and velocity (right) perturbations for models 5–8. The results are presented in Figs. 4.14 and 4.15.

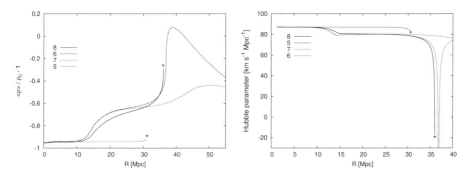

Fig. 4.14. The current average density contrast (left) and the Hubble parameter (right) for void-formation models 5–8.

equations are presented in Tables 4.2 and 4.3, and the results are shown in Figs. 4.14 and 4.15. The conclusion from numerical experiments with different shapes of the initial profiles is that a model of a void consistent with observational data (with the value of density contrast less than $\delta_0 = -0.94$, smooth edges and high density in the surrounding regions) is very hard to obtain within the L–T model, without the occurrence of a shell-crossing singularity. The final state of model 6 is very close to this singularity and in model 8 a shell crossing occurs. The best-fit model is model 6, which has both the proper density contrast and smooth edges. As before, the density fluctuations are of lesser importance. Current density profiles inside the void in models 6 and 8 are similar, but the initial amplitude of density fluctuations in model 8 is twice as large as in model 6. The main factors responsible for void formation are the velocity perturbations, with an amplitude of $\sim 8 \cdot 10^{-3}$ near the centre, and dropping below zero in the outer regions (model 7 does not fulfil this condition). Figure 4.15 presents a comparison of the average density contrasts of models 6 and 8 with the profiles derived from observation – the average profile from the North Galactic Pole region denoted

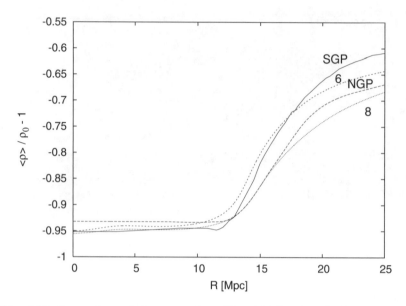

Fig. 4.15. Comparison of the curves from Fig. 4.14 (6, 8) with the observed density contrast (Hoyle and Vogeley, 2004) (SGP and NGP).

by NGP, and from the South Galactic Pole region denoted by SGP (Hoyle and Vogeley, 2004).

Since model 6 presents the best fit to observational data, let us take a closer look at its evolution. Figure 4.16 shows the density distribution at eight different moments of time. The profiles presented in Fig. 4.16 do not show the density contrast, but the real density distribution calculated from (2.3). Figure 4.17 shows respectively the functions $M(R)$ (left panel) and $E(R)$ (right panel), where R is the areal radius at the initial instant (curves 0) and at the final instant (curves 1). The pictures demonstrate the evolution of the structure: in the expanding void mass moves outwards. The amplitude of the bang time function is never larger than a few hundred years. This is negligible compared to the age of the Universe which at the moment of last scattering is of the order of 10^5 years.

4.5 Impact of radiation on void formation

In Sec. 4.4 we found that the initial amplitudes of density and velocity fluctuations needed to generate a void in an L–T model are $\sim 10^{-3}$, whereas the amplitude of baryon density fluctuations estimated from the CMB temperature fluctuations is 10^{-5}. In this section we aim to examine if this disproportion is due to matter being mostly of non-baryonic nature or to some other physical effects, like the presence of radiation. This issue was first studied by Bolejko (2006b) with the spherically symmetric Lemaître class of models described in Sec. 2.2.

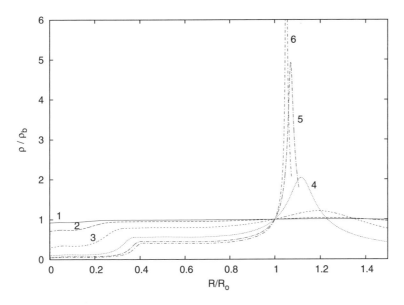

Fig. 4.16. The evolution of the density distribution in a void at ages: 10 (No. 1), 100 (No. 2) million years and 1 (No. 3), 5 (No. 4), 10 (No. 5) billion years after the Big Bang, and the current (No. 6) profile of density distribution for the model discussed in Sec. 4.4.3. R_0 is the smallest R at which the density takes the background value, and is used to mark the edge of the void.

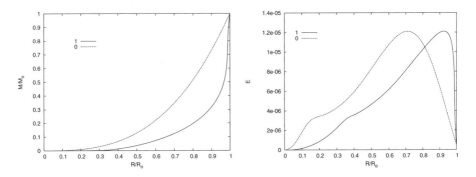

Fig. 4.17. The relative distribution of $M(R)$ (left panel) and $E(R)$ (right panel) in the void. R is the areal radius at the initial instant (curves 0) and at the final instant (curves 1).

The observed redshift of the CMB is $z \approx 1090$, and the pressure of matter at this redshift can be estimated from the perfect gas equation of state:

$$\frac{p}{\rho c^2} = \frac{k_B T}{\mu m_H c^2} \approx 10^{-10}. \tag{4.31}$$

This shows that the pressure of normal matter at last scattering is negligible.

The contribution of radiation to the energy density, however, cannot be neglected, and the ratio of radiation energy density to matter energy density at last scattering is

$$\frac{\rho_{rad}}{\rho_m} = \frac{aT^4}{\rho c^2} = \frac{\Omega_\gamma}{\Omega_m}(1+z),$$ (4.32)

which for $\Omega_m = 0.3$ gives

$$\frac{\rho_{rad}}{\rho_m} \approx 0.2.$$ (4.33)

As can be seen, this ratio decreases for low redshifts. But, at the early stage of evolution after last scattering, the radiation should have an influence on the evolution of structures. This issue will be considered further.

4.5.1 The algorithm

To calculate the evolution of a mixture of radiation and matter within the Lemaître model we first need to specify initial conditions. They consist of velocity fields of both fluids (it is assumed that matter is comoving with radiation, i.e. $u_m^\alpha = u_{rad}^\alpha = \delta^\alpha{}_0$, so that radiation contributes only to the energy-density and pressure, where the contributions obey $p_{rad} = \rho_{rad}/3$), initial fluctuations of matter density, expansion rate and radiation.

1. **Time instants**

   Since it is required that at large distances from the origin the model of a void becomes the Friedmann model, these functions are presented in the form of given fluctuations imposed on a homogeneous background. All background values are calculated for the instant of decoupling (t_1) using (4.26).

2. **The initial density perturbations**

   The radial coordinate is defined as the areal radius at the moment of last scattering, measured in kiloparsecs:

$$r = \ell = R(r, t_1)/d,$$ (4.34)

   where $d = 1$ kpc.

   The initial density fluctuations, imposed on the homogeneous background, are defined by functions of the radius ℓ, as listed in Tables 4.6 and 4.5, and the actual density fluctuations follow from

$$\rho_1(\ell) = \rho_{b1}[1 + \delta(\ell)],$$ (4.35)

   where ρ_{b1} is the density of the homogeneous background at the initial instant, and can be expressed as

$$\rho_{b1} = \rho_{b0}(1 + z_1)^3,$$ (4.36)

   where ρ_{b0} is the present value of the background density.

The mass inside the shell of radius $R(r, t_1)$ at the initial instant, measured in kiloparsecs, is again calculated from (4.28).

3. **The initial velocity perturbations**
The initial velocity perturbations imposed on the homogeneous background are defined by functions of the radius ℓ, as listed in Tables 4.6 and 4.5, and the actual velocity fluctuations follow from

$$u_1(\ell) = u_{b1}[1 + \nu(\ell)], \tag{4.37}$$

where u_{b1} is the velocity of the homogeneous background at the initial instant, and can be expressed as

$$u_{b1} = \frac{1}{c}R_{,t} = \frac{1}{c}ra_{,t}. \tag{4.38}$$

In the FLRW models the time derivative of the scale factor is given by the formula (Peebles, 1980):

$$a_{,t} = aH_0\sqrt{\mathcal{D}}, \tag{4.39}$$

where $\mathcal{D}$ is given by (4.27). Consequently, the perturbed velocity is

$$u = (1 + \nu)\frac{1}{c}RH_0\sqrt{\mathcal{D}}. \tag{4.40}$$

In Lemaître models the proper-time derivative of the areal radius R is, in consequence of the metric (2.135),

$$u = \frac{1}{c}e^{-A/2}R_{,t}, \tag{4.41}$$

from which we obtain $R_{,t}$ for a given $\nu(\ell)$.

4. **The radiation perturbations**
In general, the energy density of radiation can be written in the following form:

$$\rho_{rad}(r, t) = \rho_{rad,b}\zeta(r, t), \tag{4.42}$$

where $\rho_{rad,b}$ is the radiation energy density of a homogeneous background. Since from the last scattering instant the radiation fluctuations have always been of small amplitude, let us assume that the time dependence can be separated from the spatial dependence:

$$\zeta(r, t) = \phi(t)[1 + \gamma(r)]. \tag{4.43}$$

Let us further assume that the time-dependent amplitude of the radiation is the same as in the homogeneous background, so

$$\phi = 1. \tag{4.44}$$

However, this assumption may have to be modified in the future if observational data on the distribution of radiation become more detailed (for example, diffusion of the radiation peaks).

The radiation energy density of a homogeneous background is calculated in the usual way:

$$\rho_{rad,b} = \rho_{rad,b,0} \left(\frac{a_0}{a} \right)^4 = \rho_{rad,b,0} (1+z)^4, \tag{4.45}$$

where $\rho_{rad,b,0}$ is the present value of the radiation energy density and is equal to $4(\sigma/c)T_{CMB}^4$.

Recapitulating we have:

$$\rho_{rad} = 4\frac{\sigma}{c}T_{CMB}^4 (1+z)^4 (1+\gamma). \tag{4.46}$$

For an exact form of γ see Tables 4.6 and 4.5.

5. **Computing the evolution**

The algorithm for the evolution consists of the following steps:

(a) From (4.40), (4.41) and (2.140) the value of $R_{,t}$ and from (2.137) the value of $M_{,t}$ can be calculated. Then, using the predictor-corrector method, the value of $R(t+\tau,r)$ and $M(t+\tau,r)$ in the time step τ can be found. We further denote all the quantities found in this time step by the subscript τ.

(b) Once R_τ and M_τ are known, we can derive ρ_τ from (2.136).

(c) Then from (4.46) and the equation of state for radiation ($p = 1/3\rho_{rad}$) we derive p_τ.

(d) From ρ_τ and p_τ we can calculate A_τ by integrating (2.140).

(e) u_τ can be calculated as follows:

From (2.138) and (2.143) we obtain:

$$u^2(t,r) = \frac{2M(t,r)}{R(t,r)} + \frac{1}{3}\Lambda R^2(t,r) - 1$$

$$+ e^{-B(t_0,r)} \exp\left\{ \int_{t_0}^{t} d\tilde{t} \frac{A_{,r}(\tilde{t},r)e^{A(\tilde{t},r)/2}u(\tilde{t},r)}{R_{,r}(\tilde{t},r)} \right\}. \tag{4.47}$$

By solving this equation using the bisection method, for the time $t = t_1 + \tau$, we can calculate u_τ.

(f) Once u_τ and A_τ are known, $R_{,t}(t,r)$ and then $M_{,t}(t,r)$ can be calculated.

(g) We repeat steps (a)–(f) until t becomes the current instant.

4.5.2 Results

Measurements of the density contrast inside voids are based on observations of galaxies inside them (Hoyle and Vogeley, 2004). However, because in central regions no galaxies are observed, the real density distribution is unknown. Assuming that luminous matter is a good tracer of the total matter distribution,

Table 4.5. *Detailed descriptions of the initial fluctuations of Fig. 4.18. All the values in the table are dimensionless, and the distance parameter is the areal radius in kiloparsecs, $\ell = R_1/1$ kpc. See Table 4.6 and Figs 4.19 and 4.20.*

Profile	Parameters
$\mathcal{F}(\ell) = a \cdot f^c \cdot \mathrm{e}^{d \cdot f} + g$	$a = (A - g) \cdot (-b)^{-c} \cdot \mathrm{e}^c$ $f = \ell - b$ $g = A \cdot \left[1 - (-b)^c \cdot \mathrm{e}^{-c} \cdot h\right]^{-1}$ $h = (R - b)^c \mathrm{e}^{d(b-R)}$ $c = d \cdot b$ $d = -0.2$ $b = -4.6$ $R = 40$ $A = 6 \cdot 10^{-5}$
$\mathcal{G}(\ell) = A \cdot (b \cdot \arctan c - d \cdot \ell - \\ f \cdot \mathrm{e}^{-g^2} - \mathrm{e}^{-h^2} - \mathrm{e}^{-j^2} - \\ m \cdot \mathrm{e}^{-n^2}) \cdot k + p$	$A = -1 \cdot 10^{-4}$ $b = 4$ $c = 0.02 \cdot \ell - 0.02$ $d = \frac{5}{55}$ $f = 0.7$ $g = \ell$ $h = \frac{\ell - 1}{7}$ $j = \frac{\ell - 3}{3}$ $m = 1.225$ $n = \frac{\ell - 39}{12}$ $k = \frac{1}{1 + 0.03\ell}$ $p = -2 \cdot 10^{-5}$

and extrapolating the value of the density contrast measured on the edges of voids (where galaxies are observed) into the central regions, we can conclude that the density inside voids is below $0.06\rho_b$, which is called the limiting value. It is expected that the model will predict a present-day density inside the voids below this value.

A flat Friedmann model with $\Omega_m = 0.27$ and $\Omega_\Lambda = 0.73$ was chosen as a background model. However, in order to check how a choice of a background model affects the evolution of voids, at the end of this section three other background models will be also considered.

Figure 4.18 shows the shape of the initial perturbations, and their explicit forms are presented in Tables 4.6 and 4.5. The results plotted in Fig. 4.19 show that in four out of seven models voids are formed. As can be seen, models with an inhomogeneous distribution of radiation have no difficulty reproducing regions with density below the limiting value.

To compare our results with the observational data, Fig. 4.20 presents the average density contrast inside the voids as a function of the relative distance

Table 4.6. *Summary of run results for models of void formation with inhomogeneous radiation. See Table 4.5 and Figs. 4.18, 4.19 and 4.20.*

Run	Profile	Description
1	$\delta_- = -\mathcal{F}$ $\gamma_+ = +\frac{4}{3}\mathcal{F}$ $\nu = 1 + \mathcal{G}$	Isocurvature-like perturbations Reconstructs present-day voids
2	$\delta_- = -\mathcal{F}$ $\gamma_- = -\frac{4}{3}\mathcal{F}$ $\nu = +\mathcal{G}$	Adiabatic perturbations Collapses after 20 million years Leads to high density region
3	$\delta_- = -\mathcal{F}$ $\gamma_0 = 0$ $\nu = +\mathcal{G}$	Isothermal perturbations Does not lead to low-density region
4	$\delta_+ = +\mathcal{F}$ $\gamma_+ = +\frac{4}{3}\mathcal{F}$ $\nu = +\mathcal{G}$	Adiabatic perturbations Reconstructs the present-day voids
5	$\delta_+ = +\mathcal{F}$ $\gamma_+ = +\frac{4}{3}\mathcal{F}$ $\nu_- = -\mathcal{G}$	Reconstructs present-day voids, although density fluctuations are positive and velocity perturbations are negative
6	$\delta_0 = 0$ $\gamma_+ = +\frac{4}{3}\mathcal{F}$ $\nu_0 = 0$	Reconstructs present-day voids
7	$\delta_- = -\mathcal{F}$ $\gamma = \delta_- = -\mathcal{F}$ $\nu = +\mathcal{G}$	Leads to cluster formation, with central density $21 \cdot 10^3 \rho_b$

from the origin. The average density contrast is given by (4.29). Curve 1 presents the results of run 1, as listed in Table 4.6. Curves NGP and SGP correspond to density contrasts of voids in the 2dFGRS data estimated by Hoyle and Vogeley (2004). Although the profiles match at the centre, they do not fit accurately at the edges of voids. In our model the density contrast tends to increase faster than is observed, which could be due to our assumptions about the evolution of radiation [(4.44) and (4.45)] and would suggest that the distribution of radiation did evolve after last scattering. There is another possibility that could explain the difference between these two profiles. The density contrast estimated in Hoyle and Vogeley (2004) is based on observations of galaxies. It is possible that there is non-luminous matter within the walls, and so this procedure does not reproduce the total matter density profile sufficiently well.

Introducing radiation into the calculation we need to know the relation between matter and radiation perturbations. In linear theory there are three concepts of these relations:

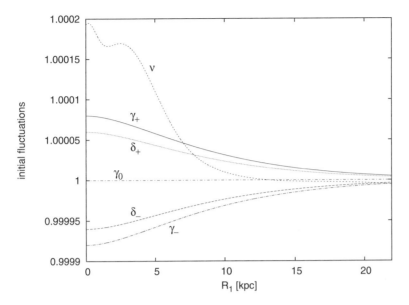

Fig. 4.18. The shape of the initial density and velocity fluctuations for void formation with inhomogeneous radiation. See Tables 4.5, 4.6 and Fig. 4.19.

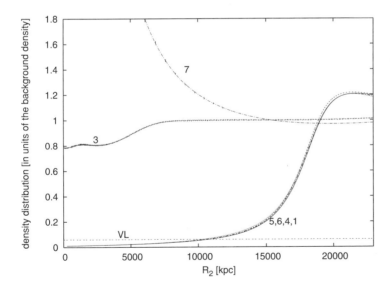

Fig. 4.19. The present-day density distribution inside voids when inhomogeneous radiation is included. The curve numbers correspond to the run numbers, as described in Table 4.6. The curve for model 2 is omitted as by t_2 it diverges near the origin. The line denoted by VL (void limit) refers to the measured density inside the voids.

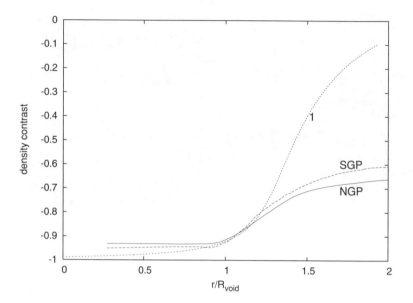

Fig. 4.20. Comparison of the present-day density contrast obtained in run 1 (curve 1) with the observed density contrast from Hoyle and Vogeley (2004) (SGP and NGP).

1. adiabatic perturbations, where $\gamma = \frac{4}{3}\delta$,
2. isocurvature perturbations, where $\gamma = \frac{4}{3}(\delta - \delta_0)$ (δ_0 is some initial value of δ),
3. isothermal perturbations, where $\gamma = 0$.

It should be stressed that in realistic conditions there are no pure adiabatic, isocurvature or isothermal perturbations and the relations between density and radiation perturbations are more complicated. However, it is instructive to know what kind of relation is more suitable for the process of void formation.

The results presented in Figure 4.19 imply that voids can be formed out of adiabatic or isocurvature perturbations and there is no significant difference between these two forms of perturbations, as long as the gradient of the radiation is negative. With isothermal perturbations, low-density regions cannot be formed as the gradient of radiation is crucial in the process of void formation.

Finally, let us notice that, in the present-day profiles of Fig. 4.20, the density gradient at the edges of voids is steeper in the models than is observed. There could be several explanations:

1. The shapes of the initial perturbations are not quite correct.
2. The assumption that the distribution of radiation did not evolve from last scattering is not fulfilled.
3. Matter around voids can distinctly depart from spherical symmetry.

4. The real density contrast increases faster than the density contrast of luminous matter.
5. The approximation involved in describing radiation as comoving with dust, with the equation of state $p_{rad} = \rho_{rad}/3$, is too crude.[9]

These points must be taken into account in future examinations. However, at this stage we can conclude that, until several million years after last scattering, radiation cannot be neglected in models of structure formation, and that the gradient of radiation is significant in the process of void formation. As was shown, the negative gradient of radiation causes faster expansion of the space inside the void, hence the density contrast decreases faster there. The excess of radiation pressure simply drives matter out of the region destined to be a void and piles it up on the edges. This effect is purely relativistic and in the model presented above radiation does not interact with matter (this is an accurate assumption for the time after the last scattering). As a result, to evolve structures like voids the amplitude of density fluctuations at last scattering does not have to be larger than 10^{-5}. Thus, the fluctuations of any non-luminous component at last scattering can be of the same amplitude as the fluctuations of baryonic matter.

Constraints on background models

The process of void formation can be used as a constraint on background models. Figure 4.21 presents the evolution of voids (with the initial conditions as in runs 1 and 3) in four different background models:

(a) $\Omega_m = 1$, $\Omega_\Lambda = 0$, $\Omega_\gamma = 4.77 \cdot 10^{-5}$,
(b) $\Omega_m = 0.4$, $\Omega_\Lambda = 0$, $\Omega_\gamma = 4.77 \cdot 10^{-5}$,
(c) $\Omega_m = 0.27$, $\Omega_\Lambda = 0$, $\Omega_\gamma = 4.77 \cdot 10^{-5}$,
(d) $\Omega_m = 0.27$, $\Omega_\Lambda = 0.73$, $\Omega_\gamma = 4.77 \cdot 10^{-5}$.

These results imply that in the absence of radiation, or of a gradient of radiation, structure formation goes on faster in the models which are filled with a greater amount of matter (curves 3a, 3b, 3c and 3d – for more details see Bolejko *et al.*, 2005). Voids cannot be formed within this kind of models.

The results in Fig. 4.21 imply that the presence of a realistic distribution of radiation is important for void formation – see models 1(c)–(d). Further, the contribution of radiation to the evolution of the system is more significant in models with smaller values of Ω_m. Therefore, the modelling of void formation can put some limits on the values of cosmological parameters. As can be seen, models with $\Omega_m \approx 0.3$ describe observed voids best. However, we cannot constrain the value of the cosmological constant in a similar manner.

---

[9] The mixture of dust and radiation should rather be described as a two-fluid distribution by the method of Haager (1997), one of the fluids being the dust, the other the radiation.

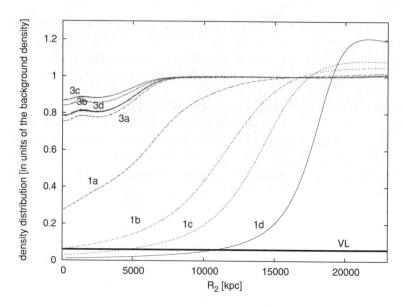

Fig. 4.21. The density distribution inside the void with three different background models. The curve numbers correspond to the run numbers (description in Table 4.6) and to the background model (as listed in the previous page). The curve denoted by VL (void limit) refers to the measured density inside voids.

4.6 Evolution of cosmic structures in different environments

In the first sections of this chapter, the spherically symmetric L–T and Lemaître models were used to study structure formation. However, the structures we observe in the Universe are far from being spherical, and we want to refine our analysis by considering a wider variety of astrophysical objects. Therefore, in this section, we investigate the evolution of small voids among compact clusters and large voids surrounded by large walls or filaments within the quasi-spherical Szekeres model.

This analysis is based on the studies by Bolejko (2006a, 2007).

4.6.1 The algorithm

To specify a Szekeres model, 5 functions of the radial coordinate need to be known. The computational algorithm used to specify the model and to calculate its evolution consists of the following steps.

1. The chosen background model is the homogeneous Friedmann model with density:

$$\rho_b = \Omega_m \times \rho_{cr} = 0.24 \times \frac{3H_0^2}{8\pi G}, \qquad (4.48)$$

where the Hubble constant is $H_0 = 74$ km s^{-1} Mpc^{-1}. The cosmological constant, Λ, corresponds to $\Omega_\Lambda = 0.76$.

2. The cosmic time of last scattering (t_1) is calculated using (4.26).
3. The radial coordinate is chosen as the value of Φ at the initial instant $t_1 = 0.5 \times 10^6$ y after the Big Bang:

$$\tilde{r} \overset{\text{def}}{=} \Phi(t_1, r). \tag{4.49}$$

For clarity in further use, the new radial coordinate is denoted r.

4. The function $M(r)$, which describes the active gravitational mass inside an $r = $ const sphere, is calculated in the following way:

$$M(r) = M_b(r) + \delta M(r), \tag{4.50}$$

where M_b is the mass in the corresponding volume of the homogeneous universe as defined in (4.28) with $\ell = r$ and $\rho = \rho_b$, and δM is a mass correction, which can be either positive or negative. The δM is defined similarly to the spherically symmetric case:

$$\delta M(r) = 4\pi \frac{G}{c^2} \int_0^r du \Phi^2(t_1, u) \Phi_{,r}(t_1, u) \delta \bar{\rho}(u), \tag{4.51}$$

where $\delta \bar{\rho}(r)$ is the function chosen to specify δM.

Although $\delta \bar{\rho}(r)$ is not the initial density fluctuation (since the latter is a function of all spatial coordinates), it gives some estimate of the initial fluctuation of the monopole density component.

5. The bang time function is assumed to be zero, i.e. $t_B(r) = 0$. Then the function $k(r)$ can be calculated from (2.149).
6. The last three functions needed to define the quasi-spherical Szekeres model are $P(r), Q(r), S(r)$. The form of these function is presented in Table 4.7
7. The evolution of the system is calculated by solving (2.146).
8. The evolution of different models is compared by their density contrast evolution. Two different types of density contrast indicators are taken into account.

The first one is the usual density contrast defined as follows:

$$\delta = \frac{\rho - \rho_b}{\rho_b}, \tag{4.52}$$

where ρ_b is the background density.

However, the density contrast defined as above is a local quantity. We can introduce another indicator of a density contrast in the way proposed by Mena and Tavakol (1999):

$$S_{IK} = \int_\Sigma \left| \frac{h^{\alpha\beta}}{\rho^I} \frac{\partial \rho}{\partial x^\alpha} \frac{\partial \rho}{\partial x^\beta} \right|^K dV, \tag{4.53}$$

where $I \in \mathbb{R}$, and $K \in \mathbb{R}\backslash\{0\}$. This family of density contrast indicators can be considered as local or global depending on the size of Σ. Such a quantity

Table 4.7. *The set of functions used to specify double structure models.*
The bang time function in the following models is assumed to be zero,
$t_B = 0$.

Model	Specification
A	$\delta\bar{\rho} = \Psi_1 e^{-(ar)^2}, \quad S = 1, \quad P = 0, \quad Q = \Phi_1 \ln(1 + lr)e^{-br}$
B	$\delta\bar{\rho} = \Psi_2 e^{-(cr)^2}, \quad S = 1, \quad P = 0, \quad Q = \Phi_2 \ln(1 + dr)e^{-br}$
C	$\delta\bar{\rho} = \Psi_1 e^{-(ar)^2}, \quad S = -(lr)^{0.4}, \quad P = \Phi_3(lr)^{0.4}, \quad Q = \Phi_4(lr)^{0.4}$
D	$\delta\bar{\rho} = \Psi_2 e^{-(cr)^2}, \quad S = -(lr)^{0.9}, \quad P = \Phi_3(lr)^{0.8}, \quad Q = \Phi_4(lr)^{0.8}$
E	$\delta\bar{\rho} = \Psi_3 e^{-(0.4ar)^2} - \Psi_4 e^{-[(lr-35)/10]^2}, \quad S = 1, \quad P = 0,$ $Q = \Phi_4(lr)^{0.8}$

Parameters

$\Psi_1 = -5 \times 10^{-3}$, $\Psi_2 = 1.14 \times 10^{-3}$, $\Psi_3 = 10^{-3}$, $\Psi_4 = 6.5 \times 10^{-4}$,
$\Phi_1 = -0.6$, $\Phi_2 = -1.45$, $\Phi_3 = 0.55$, $\Phi_4 = 0.33$, $l = 1$ kpc^{-1},
$a = 0.125$ kpc^{-1}, $b = 0.003$ kpc^{-1}, $c = 1/9$ kpc^{-1}, $d = 0.2$ kpc^{-1}

not only describes the change of density but also the change of gradients
and the volume of a perturbed region. So this density indicator describes the
evolution of the whole region in a more sophisticated way than δ. Here only
the case $I = 2, K = 1/2$ is considered.

9. The integral given by (4.53) is calculated in the quasi-spherical (ϑ, φ) coordinates of (2.151). This is because the (x, y) coordinates have an infinite range,
while in numerical calculations (all the models presented in this section are
calculated numerically) it is more convenient to use coordinates that have a
finite range.

4.6.2 Double structures

In this subsection, the evolution of double structures, namely a void with an
adjoining galaxy cluster, is investigated. Although sets of more than two structures can be described within the Szekeres model, the investigations of less complex cases are useful because they enable us to draw some general conclusions
without going into too much detail (which could easily obscure the larger picture). A more complex model is investigated in Sec. 4.6.3, and it is found that
the conclusions drawn for the double structures are still valid in such situations.

Models with $P_{,r} = 0 = S_{,r}$, $Q_{,r} \neq 0$

As mentioned above, if $P_{,r} = 0 = S_{,r} = Q_{,r}$, then the quasi-spherical Szekeres
model becomes the L–T model. Hence, the class of models considered in

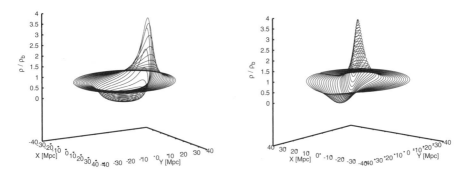

Fig. 4.22. The present-day density distribution, ρ/ρ_b of double-structure models. Left panel presents model A ($\delta M < 0$) and right panel presents model B ($\delta M > 0$).

this subsection is the simplest generalisation of the spherically symmetric models.

The double structure consisting of a void and an adjoining supercluster can be described in the Szekeres model in two different ways: with $\delta M < 0$, and with $\delta M > 0$. Both these possibilities are examined here.

Let us first consider two models – model A and B (see Table 4.7 for details). The density distributions of models A and B are presented in Fig. 4.22. As can be seen, the model with $\delta M < 0$ has a void around the centre, and the cluster is described by the high-density side of the dipole component of the matter distribution. It is the opposite in model B: the overdense region is around the origin and the void is half of the dipole component in the density.

Let us now compare the evolution of the density contrast, $\delta(t, r, x, y)$, and the $S_{2,1/2}(t, \Sigma)$ density indicator for models A and B with the corresponding models of a single void and of a single cluster obtained in the L–T model. The L–T model is considered because it can describe a single spherically symmetric structure. Such a comparison can demonstrate how the evolution of a structure changes if there is another structure in its close neighbourhood.

Figure 4.23 presents the evolution of the density contrast of model A in comparison with the corresponding model obtained in the L–T metric, which is specified by assuming the same conditions as the ones in the Szekeres model at the initial instant, namely by $t_B = 0$ and the profile of the density distribution. The local density contrast, δ, is compared at the point of the maximal and minimal density value. The left panel in Fig. 4.23 presents the evolution of the density contrast inside the void, and as can be seen, the behaviour of the density contrast in both models is similar. This is a consequence of imposing the origin conditions, where $\Phi(r_0, t) = 0 \ \forall \ t$ and some other functions are also equal to zero (for a detailed description see Sec. 2.3.3). These conditions imply that the origin behaves like a Friedmann model, and this is the reason why the quasi-spherical Szekeres and the L–T models evolve very similarly close to the origin. The right

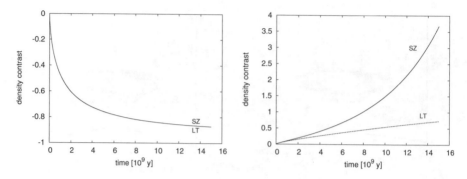

Fig. 4.23. The evolution of the density contrast, δ [Eq. (4.52)], inside the void (left panel), and inside the cluster (right panel) for model A $(\delta M < 0)$. The curve SZ presents the evolution within the Szekeres model; curve L–T presents the evolution within the Lemaître–Tolman model.

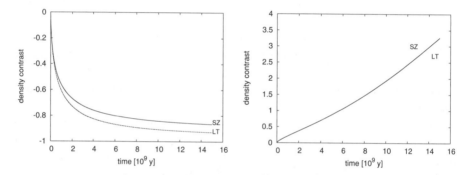

Fig. 4.24. The evolution of the density contrast, δ [Eq. (4.52)], inside the void (left panel), and inside the cluster (right panel) for model B $(\delta M > 0)$. The curve SZ presents the evolution within the Szekeres model; curve L–T presents the evolution within the Lemaître–Tolman model.

panel of Fig. 4.23 compares the evolution of the density contrast at the centre of the overdense region of model A with the corresponding L–T model, and shows the growth of the density contrast in the Szekeres model is much faster within a $\delta M < 0$ perturbation, where the mass is below the background value. This indicates that within the perturbed region of mass below the background mass $(\delta M < 0)$ the evolution of underdensities does not change, but the evolution of the overdense regions situated at the edge of the underdense regions is much faster than the corresponding evolution of isolated structures.

The evolution of the density contrast of model B $(\delta M > 0)$ is presented in Fig. 4.24. The density contrast at the point of minimal density is depicted in the left panel of Fig. 4.24, and the evolution at the origin is depicted in the right panel of Fig. 4.24. As in model A, the evolutions at the origin in the Szekeres model and in the L–T model are very similar. The evolution of the void, however, is slower within the Szekeres model than it is in the L–T model. This implies

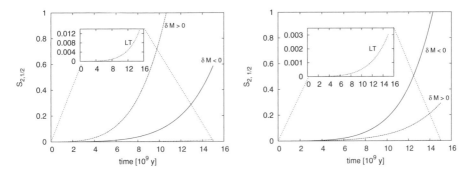

Fig. 4.25. Comparison of $S_{2,1/2}$ for the Szekeres models with $\delta M > 0$, $\delta M < 0$, and for the corresponding L–T model of a void (left panel) and of a cluster (right panel).

that single, isolated voids evolve much faster than the ones which are in the neighbourhood of large overdensities where the mass of the perturbed region is above the background mass ($\delta M > 0$).

Now, let us compare the evolution of the $S_{2,1/2}$ density indicator of (4.53). As above, two different types of Σ are considered:

A. In the case of overdense regions Σ is defined as a region where $\rho > \rho_b$.
B. In the case of underdense regions Σ is defined as a region where $\rho < \rho_b$.

Since the value of S_{IK} depends on units, the results presented in Fig. 4.25 and Fig. 4.28 are normalised so they are now of order of unity.

The left panel of Fig. 4.25 presents the evolution of $S_{2,1/2}$ for an underdense region, the right one for an overdense region. As can be seen, $S_{2,1/2}$ for the two Szekeres models are comparable and the growth of $S_{2,1/2}$ for the L–T model is much slower. This is because the volumes of the considered regions are different. In the Szekeres model the volume is larger than the volume in the L–T model.

Figure 4.22 presents the shape of the structures without corrections for the shell displacement. For example, the void in Fig. 4.22 (left panel) and in Fig. 4.26 seems to be almost spherical. In fact this void is squeezed in the $+Y$ direction and elongated in the $-Y$ direction [$Q_{,r} \neq 0$, $P_{,r} = 0 = S_{,r}$ – this follows from the metric (2.150) and Eqs. (2.164), (2.151)]. This also leads in some regions to density gradients larger than in the L–T model, which causes such a large disproportion in $S_{2,1/2}$ between the Szekeres and L–T models.

The results presented above indicate that the evolution of a Szekeres model is much more complex than that of an L–T model. The evolution not only depends on the amplitude of the density contrast, but also on the density gradients and on the volume of the perturbed region. This is the reason why the $S_{2,1/2}$ curve for the void in model B ($\delta M > 0$) is higher than in other models, although the density contrast in this model evolves more slowly than in model A. Similarly, as can be seen by comparison of Figs. 4.26 and 4.27, the overdense region in the

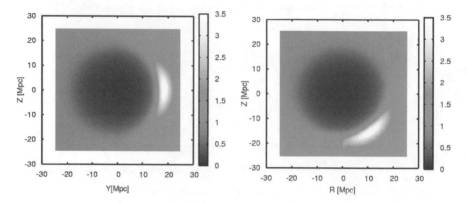

Fig. 4.26. The present-day colour-coded density distribution, ρ/ρ_b, of double-structure models A and C (in which $\delta M < 0$). Left panel – $P_{,r} = S_{,r} = 0$. Right panel – $P_{,r} \neq 0 \neq S_{,r}$. Slices through the origin and dipole axis are shown (on the left panel the dipole axis coincides with the Y-axis; on the right panel it is $R = \sqrt{X^2 + Y^2} = 0$). White indicates a high-density region.

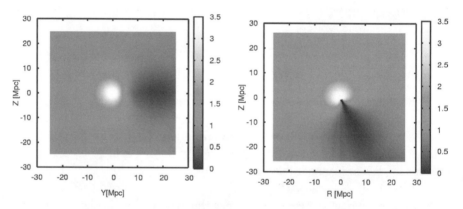

Fig. 4.27. The present-day colour-coded density distribution, ρ/ρ_b, of double-structure models B and D (in which $\delta M > 0$). Left panel – $P_{,r} = S_{,r} = 0$. Right panel – $P_{,r} \neq 0 \neq S_{,r}$.

model with $\delta M < 0$ is much larger than in other models, and as a consequence the $S_{2,1/2}$ value for this model evolves much faster than in other models. The $S_{2,1/2}$ plots give us information about the evolution of the whole perturbed region.

The evolution of the density at a single point is described by the local density contrast δ. As can be seen, the evolution of the maximal and minimal density contrast depends on the value of δM in the perturbed region. The evolution of the density contrast inside large and isolated voids is faster than inside small voids which are surrounded by highly dense regions. On the other hand, the evolution of the density contrast in high-density regions in close neighbourhood of large voids is faster, due to faster mass flow from the voids.

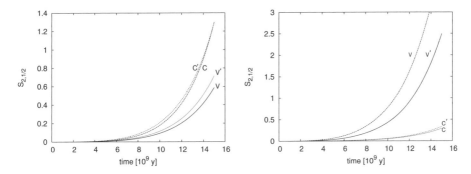

Fig. 4.28. Comparison of $S_{2,1/2}$ for double-structure models with $\delta M < 0$ (left panel) and with $\delta M > 0$ (right panel). C corresponds to a 'cluster' – an overdense region, and V corresponds to a 'void' – underdense region. Primes denote models with $P_{,r} \neq 0 \neq S_{,r}$.

Models with $P_{,r} \neq 0 \neq S_{,r}$, $Q_{,r} \neq 0$

Now let us consider two models of non-constant P, Q and S – models C and D (see Table 4.7 for details).

Figure 4.26 presents the comparison of the present-day density distribution in models A and C in colour-coded diagrams. It presents the cross sections perpendicular to the XY surface of the considered structures. The left panel of Fig. 4.26 presents the cross section through the surface $\phi = \pi/2$ and the right panel presents the cross section through the surface $\phi \approx \pi/6$. Figure 4.27 presents the cross sections of models B and D. As can be seen, each pair of structures appears to be similar but, in comparison with model A, the void in model C is moved down and right. In model D, on the other hand, it is the cluster which is moved down and right as compared to model B.

The evolutions of the density contrasts inside the voids and clusters of models C and A are very similar, which should not be surprising, as model C has the same $\bar{\delta}(r)$ as model A. Also, the evolutions of the corresponding density contrasts of models D and B are similar. The functions (S, P, Q) were chosen so that they reproduce the same shapes of current structures and the same density contrasts inside them. However, it is not clear whether the evolutions of $S_{2,1/2}$ are comparable, too. When the functions (S, P, Q) are not constant, the axis of the mass-dipole changes from one sphere to another.

Also, volumes of the perturbed regions and the density gradients can be different. So it may be interesting to compare the evolution of the whole perturbed underdense and overdense regions of models A, B, C and D.

Figure 4.28 presents the comparison of the $S_{2,1/2}$ evolution of models A–D. The primed letters denote models with $S_{,r} \neq 0 \neq P_{,r}$, $Q_{,r} \neq 0$. As can be seen, the evolution of $S_{2,1/2}$ for all these models is also comparable. These results confirm what might have been intuitively guessed, i.e. that the evolution in the quasi-spherical Szekeres model does not depend on the position of the dipole component. As long as the shape and the density contrast of the models are similar, they evolve in a very similar way.

4.6.3 Connection to the large-scale structure of the Universe

There remains the problem whether the conclusions presented in the previous subsection are not limited to the class of models considered in this book. Are they general? How relevant are they for the real large-scale structure of the Universe? These questions are addressed in this subsection.

First of all, let us consider whether the choice of functions used so far does not significantly restrict our analysis. The models presented above were defined by choosing (t_B, M, S, Q, P). As can be seen, the functions (S, Q, P) describe the position of the dipole and even with S and P constant, we are still able to reconstruct the cosmic structures. The other functions which specify the model are M and t_B. It is also possible to choose other sets of functions, such as k, and t_B, or any k and M. In practice, however, we cannot take any arbitrary k and M (or any other combination along the lines described for L–T models in Sec. 2.1.5) because such arbitrary choices may lead to shell crossings or to fluctuations at last scattering that are too large, and in most cases to both of these. Typically, fluctuations in the bang time t_B, which generate decaying modes (Silk, 1977), should be less than a few thousand years to satisfy CMB constraints, though carefully chosen combinations of large-amplitude fluctuations in these three arbitrary functions may be possible. However, one does not expect large differences in the age of the Universe within the region of size of several to several tens of Mpc. So if t_B is set (say that in the studied region $t_B \approx 0$), then we are left with one function to manipulate with. Let it be M, or, more intuitively, δM. As seen from (2.149), if δM increases, then k must decrease in order to preserve the constant age of the Universe. On the other hand, the decrease of k leads to the decrease of the expansion rate, θ (2.153). This, further, via the continuity equation (2.154), influences the evolution. If we look at the distribution of the expansion rate, which is presented in Fig. 4.29, we can see that the expansion rate closely follows the pattern of the density distribution (Figs. 4.26 and 4.27). Figure 4.29 presents the ratio θ_{SZ}/θ_b of the expansion parameter in the Szekeres models (2.153) to the expansion parameter in the homogeneous background.

Now it should be clear why the void in models A and C ($\delta M < 0$) evolves much faster than in models B and D ($\delta M > 0$). This is because if the mass of the perturbed region is below the background mass, such a region expands much faster than the background, leading to the formation of large underdense regions. Let us also notice that similar patterns are present in more complicated configurations. Let us now consider model E which describes the evolution of a triple structure. All functions used to specify this model are presented in Table 4.7. Figure 4.30 presents the density distribution with two overdense regions, one at the origin, and a small void between them which extends round the centre at

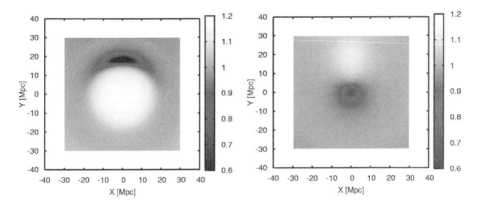

Fig. 4.29. The ratio θ_{SZ}/θ_b of the expansion scalar in the model to that of the background, at the current instant, showing a slice through the origin. White indicates a rapid expansion region. Left panel presents the ratio in model A, right panel presents the ratio in model B.

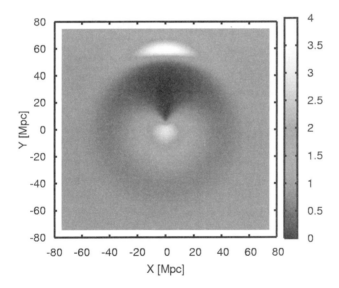

Fig. 4.30. The present-day colour-coded density distribution, ρ/ρ_b, for the triple structure of model E, showing a slice through the origin. White indicates a high-density region.

larger radii. The evolution of model E is presented in Fig. 4.31 for 5 different time instants. For clarity, the plotted profiles are those along the line $X = 0$ in Fig. 4.30. As can be seen, the part which is between overdensities evolves faster in regions where the void is wider than in regions close to the origin where it is narrower. Another significant fact is that the overdense region, which interfaces with the void across a larger area, evolves much faster than the cluster at the origin, which is more compact. This model exhibits the features of the models

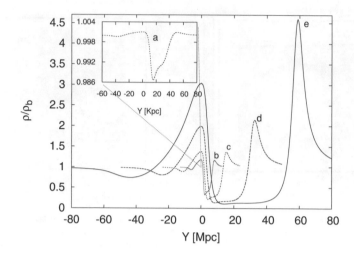

Fig. 4.31. The evolution of the density profile of the triple structure in model E. The profiles correspond to the line $X = 0$ in Fig. 4.30. Letters correspond to different time instants: a – 0.5×10^6 y after the Big Bang; b – 1.5×10^9 y; c – 5×10^9 y; d – 10×10^9 y; e – present instant.

previously considered. Thus, it might be speculated that the evolution of real structures follows similar patterns. Namely, small voids in the Universe which are surrounded by large high-density regions evolve more slowly than large isolated voids. From the perspective of the continuity equation the expansion of the space in this region is very slow and this is the reason why the voids do not evolve as fast as they otherwise could. On the other hand, the expansion is much faster inside large voids, where the mass of the perturbed region is below the background mass ($\delta M < 0$). In such structures matter flows from voids towards the dense regions which form at their larger sides and enhance their evolution. In this way, the higher expansion rate inside large voids leads to the formation of large and elongated structures such as walls and filaments which emerge at their edges.

5

The cosmological constant and coincidence problems

Since its discovery during the late 1990s (Riess *et al.*, 1998; Perlmutter *et al.*, 1999), the dimming of distant SN Ia has been mostly ascribed to the influence of a mysterious dark energy component. Formulated in a Friedmannian framework, based upon the 'cosmological principle', this interpretation has given rise to the 'Concordance' model. However, we have already stressed in Chapter 1 that a caveat of such a reasoning is that the 'cosmological principle' derives from the philosophical Copernican assumption and has never been tested.

Moreover, it is well known, since the work of Ellis and Stoeger (1987), that the inhomogeneities observed in our Universe can have an effect upon the values of the cosmological parameters derived for a smoothed-out or averaged Friedmannian model. A tentative estimate of such an effect was computed by Hellaby (1988). He found that the error obtained when using averaging procedures compared to the volume matching (i.e. the procedure proposed by Ellis and Stoeger, 1987) of Friedmann models to inhomogeneous L–T solutions with realistic density profiles implies that the mean density and pressure of the averaged Friedmann models are 10–30% underestimated as regards the volume-matched ones. Therefore, even if the cosmological constant is the only component in the Einstein equations beyond ordinary matter, the estimation of its actual value is less straightforward than the conventional wisdom has it.

As regards dark energy, i.e. a negative pressure fluid with an equation of state parameter $w \neq -1$, ($w = -1$ being the signature of the pure geometric cosmological constant), it remains a phenomenon which cannot be explained in the framework of current physics. This is known as 'the cosmological constant problem'.

Another feature of the luminosity distance–redshift relation inferred from the supernova data when they are analysed in a Friedmannian framework is that it yields a late-time acceleration of the expansion rate, beginning close to the epoch when structure formation enters the nonlinear regime. This would imply that we live at a time when the matter energy density and the dark energy are of the same order of magnitude. This is known as 'the coincidence problem'.

To deal with these problems some authors have proposed a modification of the general relativity theory at large distance scales (see e.g. Carroll *et al.*, 2004; Capozziello *et al.*, 2006).

However, our current purpose is to describe work dedicated to a simpler and more natural proposal, which only makes use of known physics and phenomena. Since it appears that the onset of apparent acceleration and the beginning of structure formation in the Universe are concomitant, the idea that the SN Ia observations could be reproduced totally or partially by the effect of large-scale inhomogeneities was put forward. This interpretation was first proposed independently by a few authors (Dąbrowski and Hendry, 1998; Pascual-Sánchez, 1999; Célérier, 2000a; Tomita, 2000), shortly after the release of the data. Then, after a period of relative disaffection, it experienced a renewed interest.

Three different methods have been used to deal with this proposal: two back-reaction computations, one using an averaging procedure, the other within a perturbative scheme, and the use of exact inhomogeneous models, in particular those involving L–T solutions (see Célérier, 2007a, for a review). Since the purpose of the present book is to study how relativistic cosmology can be done by exact methods, we shall focus our attention on the analyses completed within an L–T framework.

Note that another explanation of the observed dimming of the SN Ia is that it might be due to the effect of an actual geometrical cosmological constant. Its value such as appears in the ΛCDM model was predicted in the framework of scale relativity, as $\Lambda = 1.36 \times 10^{-56}$ cm^{-2}, long before the release of the supernova results (Nottale, 1993 and 1996). However, that approach is beyond the scope of this book.

5.1 Constraints on model parameters from the SN Ia data

This section summarises part of some work done by Célérier (2000a) proposing a test of the cosmological 'principle' with the SN Ia observations and studying the possibility of explaining the cosmological constant 'problem' as an effect of large-scale inhomogeneities.

The use of SN Ia as 'standard candles' to probe consistently the geometry of the Universe depends on the robustness of the method by which every source of potential bias or systematic uncertainties is correctly taken into account. An evolution of the progenitors is generally considered as one of the most likely sources of potential systematic errors in the analysis of the supernova data. Mustapha *et al.* (1997) have made the point that, given isotropic observations about us, homogeneity cannot be proved without either a model-independent theory of source evolution, or distance measures that are independent of source evolution. Therefore, should a perceptible SN Ia evolution effect be demonstrated, its impact would have to be evaluated with a model-independent method. Other main assumptions are: dust extinction can be neglected and there are no mechanisms for photon decay. We will not discuss here these important issues, since they are beyond the scope of the present book, and we will adopt the 'standard candle' assumption while keeping in mind the fate of previously claimed 'standard candles', such as, e.g. Cepheid variables. Remember that, in the 1940s, Baade (1956) discovered that the nearby Cepheid variables used to calibrate the standard candle were population I stars with metal content much higher than the distant population II stars used to measure distances to nearby galaxies. Therefore, the period–luminosity relation of the Cepheids had to be revised for the second time (the first time was when the importance of interstellar extinction was realised).

The luminosity distance D_L of a source is defined as the distance from which the radiating body, if motionless in a Euclidean space, would produce an energy flux equal to the one measured by the observer. It follows therefore that

$$l = \frac{L}{4\pi D_L^2},$$ (5.1)

L being the absolute luminosity, i.e. the emitted power in the rest frame of the source, and l, the measured bolometric flux, i.e. integrated over all frequencies by the observer.

For a given isotropic cosmological model the distance at redshift z is a function of z and of the model parameters. The apparent bolometric magnitude m of a standard candle of absolute bolometric magnitude M is also a function of z and of the parameters of the model. It can be written, for D_L in megaparsecs, as

$$m = M + 5 \log D_L(z; cp) + 25,$$ (5.2)

where M is the magnitude measured at $D_{10} = 10$ pc,

$$M = 2.5 \log \frac{L}{4\pi D_{10}^2}.$$ (5.3)

Here, cp denotes the set of cosmological parameters pertaining to the given model, which can be either constants, as in FLRW models, or functions of z, as in L–T ones. The diameter or area distance D_A is also defined by the formula one would use in Euclidean space,

$$D_A = \frac{D}{\alpha},$$ (5.4)

where α is the measured angular diameter of a source and D is its physical diameter. The reciprocity theorem (Etherington, 1933; Penrose, 1966; Ellis, 1971; Plebański and Krasiński, 2006; see Ellis, 2007) relates these two distances,

$$D_L = (1 + z)^2 D_A.$$ (5.5)

If one considers any cosmological model for which the luminosity distance D_L is a function of the redshift z and of the cosmological parameters cp, and if this function is Taylor expandable near the observer, i.e. for $z < 1$, then

$$D_L(z; cp) = \left(\frac{\mathrm{d}D_L}{\mathrm{d}z}\right)_{z=0} z + \frac{1}{2}\left(\frac{\mathrm{d}^2 D_L}{\mathrm{d}z^2}\right)_{z=0} z^2$$

$$+ \frac{1}{6}\left(\frac{\mathrm{d}^3 D_L}{\mathrm{d}z^3}\right)_{z=0} z^3 + \frac{1}{24}\left(\frac{\mathrm{d}^4 D_L}{\mathrm{d}z^4}\right)_{z=0} z^4 + \mathcal{O}(z^5), \qquad (5.6)$$

since $D_L(z = 0) = 0$.

Luminosity distance measurements of sources at different redshifts $z < 1$ yield values for the different coefficients in the above expansion. Going to higher red-shifts amounts to measuring the coefficients of higher powers of z. For very low redshifts, the leading term is first order; for intermediate redshifts, second order. Then third-order terms provide significant contributions. For redshifts approaching unity, higher-order terms can no longer be neglected.[1]

Therefore, for cosmological models with a very large number of free param-eters, or with parameters which are not constants but, for example, functions of the redshift, such measurements only provide constraints upon the values of the parameters near the observer. But for cosmological models with few param-eters, giving independent contributions to each coefficient in the expansion, the method not only provides a way to evaluate these parameters, but, in most cases, to test the validity of the model itself by checking the constraints on the Taylor coefficients. These can be derived with methods such as the one below, developed for FLRW cosmologies.

5.1.1 The ΛCDM cosmological model

For FLRW models with a cosmological constant, the luminosity distance is (Carroll *et al.*, 1992)

$$D_L = \frac{c(1+z)}{H_0\sqrt{|k|}}\, \mathcal{S}\left(\sqrt{|k|}\int_0^z [(1+z')^2(1+\Omega_m z') - z'(2+z')\Omega_\Lambda]^{-1/2}\mathrm{d}z'\right),$$

$$(5.7)$$

H_0 being the current Hubble constant, Ω_m, the total mass density parameter, Ω_Λ, the cosmological constant parameter, and

$$\mathcal{S} = \sin \quad \text{and} \quad k = 1 - \Omega_m - \Omega_\Lambda \quad \text{for} \quad \Omega_m + \Omega_\Lambda > 1,$$
$$\mathcal{S} = \sinh \quad \text{and} \quad k = 1 - \Omega_m - \Omega_\Lambda \quad \text{for} \quad \Omega_m + \Omega_\Lambda < 1,$$
$$\mathcal{S} = I \quad\quad \text{and} \quad k = 1 \quad\quad\quad\quad\quad\quad \text{for} \quad \Omega_m + \Omega_\Lambda = 1,$$

I being the identity operator.

---

[1] In fact, at variance with what was advocated in (Célérier, 2000a), (5.6) does not actually apply for $z < 1$ but for $z < z_0$, where z_0 is the radius of convergence of the Taylor series. The z_0, which can be smaller than 1, is different for each function $D_L(z)$ and has to be determined separately for each cosmological model. However, since the conclusions of the following reasoning do not depend on the z_0 value and since $z_0 = 1$ applies to most $D_L(z)$ functions, we shall retain such a value in the subsequent developments.

The Taylor expansion of (5.6) applied to this expression for the FLRW luminosity distance yields

$$D_L^{(1)} \equiv \left(\frac{\mathrm{d}D_L}{\mathrm{d}z}\right)_{z=0} = \frac{c}{H_0}, \tag{5.8}$$

$$D_L^{(2)} \equiv \frac{1}{2}\left(\frac{\mathrm{d}^2 D_L}{\mathrm{d}z^2}\right)_{z=0} = \frac{c}{4H_0}(2 - \Omega_m + 2\Omega_\Lambda), \tag{5.9}$$

$$D_L^{(3)} \equiv \frac{1}{6}\left(\frac{\mathrm{d}^3 D_L}{\mathrm{d}z^3}\right)_{z=0} = \frac{c}{8H_0}(-2\Omega_m - 4\Omega_\Lambda - 4\Omega_m\Omega_\Lambda$$
$$+ \Omega_m^2 + 4\Omega_\Lambda^2), \tag{5.10}$$

$$D_L^{(4)} \equiv \frac{1}{24}\left(\frac{\mathrm{d}^4 D_L}{\mathrm{d}z^4}\right)_{z=0} = \frac{5c}{72H_0}(8\Omega_\Lambda + 4\Omega_m\Omega_\Lambda + 2\Omega_m^2$$
$$- 16\Omega_\Lambda^2 - 12\Omega_m\Omega_\Lambda^2 + 6\Omega_m^2\Omega_\Lambda - \Omega_m^3 + 8\Omega_\Lambda^3). \tag{5.11}$$

These coefficients are independent functions of the three parameters of this class of models, H_0, Ω_m and Ω_Λ.

Now, a simplified way of dealing with the SN Ia data in the framework of a Friedmannian cosmology is to use the magnitude 'zero-point' method. The magnitude 'zero-point' $\mathcal{M} \equiv M - 5\log H_0 + 25$ can be measured from the apparent magnitude and redshift of low-redshift examples of the standard candles, without knowing H_0, since, in this representation, the apparent magnitude is

$$m \equiv \mathcal{M} + 5\log \mathcal{D}_L(z; \Omega_m, \Omega_\Lambda). \tag{5.12}$$

Since $\mathcal{D}_L(z; \Omega_m, \Omega_\Lambda) \equiv H_0 D_L(z; \Omega_m, \Omega_\Lambda, H_0)$, it depends on Ω_m and Ω_Λ alone and such is the case for its expansion coefficients $\mathcal{D}_L^{(i)}$. We can therefore derive conditions which must be fulfilled by these coefficients when fitting them to the data. These conditions are mandatory for any ΛCDM model to represent our observed Universe.

In the standard CDM model, i.e. for $\Omega_\Lambda = 0$,

$$\mathcal{D}_L^{(2)} = \frac{c}{4}(2 - \Omega_m). \tag{5.13}$$

If the analysis of the measurements gives $\mathcal{D}_L^{(1)} < 2\mathcal{D}_L^{(2)}$, it implies $\Omega_m < 0$, which is physically irrelevant. The model is therefore ruled out. This is what happens with the SN Ia data, and what induced the postulate of a strictly positive cosmological constant, to counteract the $-\Omega_m$ term.

Now, to test FLRW models with $\Omega_\Lambda \neq 0$, one has to go at least to the third order to have any chance of obtaining a result. A ruling out of these models at the third-order level, i.e. due to a negative value for Ω_m, would occur if it were found that (see (5.8)–(5.10))

$$1 - \frac{\mathcal{D}_L^{(3)}}{\mathcal{D}_L^{(1)}} - 3\frac{\mathcal{D}_L^{(2)}}{\mathcal{D}_L^{(1)}} + 2\left(\frac{\mathcal{D}_L^{(2)}}{\mathcal{D}_L^{(1)}}\right)^2 < 0. \tag{5.14}$$

A final test for the homogeneity hypothesis, on the part of our past light cone spanned by the supernova data, would be a check of the necessary condition, obtained from (5.8)–(5.11), which can be written as

$$\mathcal{D}_L^{(4)} = -\frac{10}{9}\,\mathcal{D}_L^{(2)} + \frac{10}{3}\,\frac{\mathcal{D}_L^{(2)2}}{\mathcal{D}_L^{(1)}} - \frac{20}{9}\,\frac{\mathcal{D}_L^{(2)3}}{\mathcal{D}_L^{(1)2}} - \frac{20}{9}\,\mathcal{D}_L^{(3)} + \frac{10}{3}\,\frac{\mathcal{D}_L^{(2)}\mathcal{D}_L^{(3)}}{\mathcal{D}_L^{(1)}}. \qquad (5.15)$$

It must be stressed here that the test described above only applies to data from $z < 1$ supernovae. Although ongoing surveys are extending to redshifts higher than unity, where the Taylor expansion is no longer valid, the above tests might be performed with subsamples of $z < 1$ sources.

Now, one could argue that best-fit confidence regions in parameter space, such as those published in the literature to present the results of the data analysis, could be a way of dealing with this problem. If these regions were located in physically irrelevant domains of the parameter space, e.g. $\Omega_m < 0$, this would rule out the Friedmannian models as able to reproduce the geometry of our local Universe. It must be stressed, however, that the computation of these best-fit confidence regions proceeds from a Bayesian data analysis, for which a prior probability distribution accounting for the physically (in Friedmann cosmology) allowed part of the parameter space is assumed. The results are thus distorted by an a priori homogeneity assumption, which would have to be discarded for the completion of any robust test of this hypothesis.

Note also that, even though we have only considered here a ΛCDM model, any other parameters can be taken into account following the above scheme. As an example, a constant equation of state parameter $w \neq -1$ can be added in the expression for the luminosity distance and its contribution to the Taylor expansion will result in a new condition for the coefficients. A redshift-dependent $w(z)$ might also be included in the form of its Taylor expansion up to whatever order would provide no more information, since the higher-order terms might become of the same magnitude as the observational errors.

5.1.2 L–T models with zero cosmological constant

We examine in this section the possibility of reproducing the SN Ia data in an L–T model with no cosmological constant, and in the case where the observer is assumed to be located near the symmetry centre of the model.

The complexity of the redshift–distance relation in inhomogeneous models and their deviation from the FLRW relation were stressed by Kurki-Suonio and Liang (1992) and Mustapha *et al.* (1998). In an earlier very interesting paper, Partovi and Mashhoon (1984) showed that the luminosity distance–redshift relation in local models with radial inhomogeneities and the barotropic equation of state for the source cannot be distinguished from the FLRW one, at least to second order in z.

Since the L–T solution appears to be a good tool for the study of the observed Universe in the matter-dominated era, it is used here as an example showing that a non-vanishing cosmological constant in a Friedmann universe can be replaced, under some conditions upon the functions of r defining each model, by inhomogeneity with a zero cosmological constant, and is able to fit the SN Ia data just as well.

In the following, we set $t_B(0) = 0$ at the symmetry centre $(r = 0)$ by an appropriate translation of the $t = $ const surfaces, and describe the Universe by the $t > t_B(r)$ part of the (r, t) plane, increasing t corresponding to going from the past to the future.

When evaluated on the past null cone of a central observer, the areal radius $R_n = R(t_n(r), r)$ is precisely the diameter distance[2]

$$D_A = R_n. \tag{5.16}$$

From the reciprocity theorem (5.5) and (5.16) we have

$$D_L = (1 + z)^2 R_n, \tag{5.17}$$

and one sees that the luminosity distance D_L is a function of the redshift z and, through R_n, of the parameter functions of the model: $M(r)$, $E(r)$ and $t_B(r)$. Successive partial derivatives of R with respect to r and t, and derivatives of $E(r)$ with respect to r, evaluated at the observer, contribute to the expressions for the coefficients of the luminosity distance expansion of (5.17) in powers of z. It is therefore interesting to note the behaviour of $R(t, r)$ and $E(r)$ near the symmetry centre of the model, i.e. near the observer, which takes the form (see e.g. Humphreys *et al.*, 1997, r is defined by $2M(r) = r^3$)

$$R(t, r) = R_{,r}(t, 0) r + \mathcal{O}(r^2), \tag{5.18}$$

$$E(r) = \frac{1}{2} E_{,rr}(0) r^2 + \mathcal{O}(r^3), \tag{5.19}$$

thanks to the regularity conditions at the centre which imply $R(t, 0) = R_{,t}(t, 0) = E(0) = E_{,r}(0) = 0$, as well as the vanishing of higher-order derivatives of R with respect to t alone at $r = 0$.

The expressions for the coefficients of the luminosity distance expansion naturally follow. After some calculations, one obtains

$$D_L^{(1)} = \frac{R_{,r}}{R_{,tr}}, \tag{5.20}$$

$$D_L^{(2)} = \frac{1}{2} \frac{R_{,r}}{R_{,tr}} \left(1 + \frac{R_{,r} R_{,ttr}}{R_{,tr}^2} + \frac{R_{,rr}}{R_{,r} R_{,tr}} - \frac{R_{,trr}}{R_{,tr}^2} \right), \tag{5.21}$$

---

[2] Bondi (1947) presents it as the radial luminosity distance of a source, but with our definitions this is not correct. However, some authors mean 'corrected luminosity distance' when they write 'luminosity distance'.

$$D_L^{(3)} = \frac{1}{6} \frac{R_{,r}}{R_{,tr}} \left(-1 - \frac{R_{,r} R_{,ttr}}{R_{,tr}^2} + 3 \frac{R_{,r}^2 R_{,ttr}^2}{R_{,tr}^4} - \frac{R_{,r}^2 R_{,tttr}}{R_{,tr}^3} \right.$$
$$- 6 \frac{R_{,r} R_{,ttr} R_{,trr}}{R_{,tr}^4} + 4 \frac{R_{,rr} R_{,ttr}}{R_{,tr}^3} + 2 \frac{R_{,r} R_{,ttrr}}{R_{,tr}^3} - 3 \frac{R_{,rr} R_{,trr}}{R_{,r} R_{,tr}^3}$$
$$\left. + 3 \frac{R_{,trr}^2}{R_{,tr}^4} + \frac{R_{,rrr}}{R_{,r} R_{,tr}^2} - \frac{R_{,trrr}}{R_{,tr}^3} + \frac{E_{,rr}}{R_{,tr}^2} \right), \tag{5.22}$$

with implicit evaluation of the partial derivatives at the observer.

To compare the expansions of D_L in the FLRW and in the L–T cases, we recall that the expressions for the Hubble parameter, H_0, and of the deceleration parameter, q_0, at the observer in the FLRW model are given, in units in which $c = 1$, by

$$H_0 = \frac{1}{D_L^{(1)}}, \tag{5.23}$$

$$q_0 = 1 - 2H_0 D_L^{(2)}. \tag{5.24}$$

The SN Ia data show the q_0 of (5.24) is negative, and it is claimed that this supports the Concordance model.

Substituting these formulae as well as (5.8)–(5.10) into (5.20)–(5.22), it is easy to see that the expressions for the FLRW coefficients in the expansion of D_L in powers of z can mimic L–T ($\Lambda = 0$) ones, at least to third order. This is straightforward for $D_L^{(1)}$. The case of $D_L^{(2)}$ is discussed at length in Partovi and Mashhoon (1984). For higher-order terms, it implies constraints on the L–T parameters, which will be illustrated below with the particular example of flat models. In fact, owing to the appearance of higher-order derivatives of the functions of the parameter η in each higher-order coefficient, L–T models are completely degenerate with respect to any given magnitude–redshift relation in the sense that one of its arbitrary functions remains unspecified. In contrast, FLRW models with constant parameters, including the equation of state parameter w, are more rapidly constrained and therefore cannot fit any given relation when tested at sufficiently high order in the expansion.

5.1.3 *Illustration: flat L–T* ($\Lambda = 0$) *models*

To illustrate the kind of constraints that can be imposed on L–T parameters by observational data, the particular case of spatially flat L–T ($E = \Lambda = 0$) models is analysed here.

In this case, the expression for R is given by (2.6) and the calculation of its successive derivatives, contributing to the expressions of the expansion coefficients, is straightforward.

The mass function $M(r)$ can be used to define the radial coordinate r: $M(r) \equiv M_0 r^3$, where M_0 is a constant.

With the covariant definition for H_0 mentioned in (5.23), the $\mathcal{D}_L^{(i)}$'s, as derived from (5.20) to (5.22), can be written, in units in which $c = 1$, as

$$D_L^{(1)} = \frac{1}{H_0}, \tag{5.25}$$

$$D_L^{(2)} = \frac{1}{4H_0}\left(1 - 6\frac{t_{B,r}(0)}{(9GM_0/2)^{1/3}\,t_p^{\,2/3}}\right), \tag{5.26}$$

$$D_L^{(3)} = \frac{1}{8H_0}\left(-1 + 4\frac{t_{B,r}(0)}{(9GM_0/2)^{1/3}\,t_p^{\,2/3}} + 6\frac{t_{B,r}^{\,2}(0)}{(9GM_0/2)^{2/3}\,t_p^{\,4/3}}\right.$$
$$\left. - 9\frac{t_{B,rr}(0)}{(9GM_0/2)^{2/3}\,t_p^{\,1/3}}\right), \tag{5.27}$$

with the previously indicated choice $t_B(0) = 0$, and where t_p is the time coordinate at the observer. It is convenient to note that t_p is not a free parameter of the model, since its value follows from the currently measured temperature at 2.73 K (Célérier and Schneider, 1998). Note also the slight changes in the numerical coefficients of (5.27) which were corrected from their values published in Célérier (2000a).

A comparison with the corresponding FLRW coefficients (5.8) to (5.10) gives the following relations:

$$\Omega_m \longleftrightarrow 1 + 5\,\frac{t_{B,r}(0)}{(9GM_0/2)^{1/3}\,t_p^{\,2/3}} + \frac{15}{2}\,\frac{t_{B,r}^2(0)}{(9GM_0/2)^{2/3}\,t_p^{\,4/3}}$$
$$+ \frac{9}{4}\,\frac{t_{B,rr}(0)}{(9GM_0/2)^{2/3}\,t_p^{\,1/3}}, \tag{5.28}$$

$$\Omega_\Lambda \longleftrightarrow -\frac{1}{2}\,\frac{t_{B,r}(0)}{(9GM_0/2)^{1/3}\,t_p^{\,2/3}} + \frac{15}{4}\,\frac{t_{B,r}^2(0)}{(9GM_0/2)^{2/3}\,t_p^{\,4/3}}$$
$$+ \frac{9}{8}\,\frac{t_{B,rr}(0)}{(9GM_0/2)^{2/3}\,t_p^{\,1/3}}. \tag{5.29}$$

Equation (5.29) implies that a nonvanishing cosmological constant in an FLRW interpretation of the data at $z < 1$ corresponds to a mere constraint on the model parameters in a flat L–T ($\Lambda = 0$) interpretation.

Therefore, any magnitude–redshift relation established up to the third-order level, can be interpreted in either model. For instance, the results published in Perlmutter *et al.* (1999) under the following form pertaining to an FLRW interpretation

$$0.8\,\Omega_m - 0.6\,\Omega_\Lambda \approx -0.2 \pm 0.1, \tag{5.30}$$

should be written, with no model-dependent a priori assumption, as

$$2 \left(\frac{\mathcal{D}_L^{(2)}}{\mathcal{D}_L^{(1)}} \right)^2 - 4.2 \frac{\mathcal{D}_L^{(2)}}{\mathcal{D}_L^{(1)}} - \frac{\mathcal{D}_L^{(3)}}{\mathcal{D}_L^{(1)}} \approx -1.8 \pm 0.1, \tag{5.31}$$

and it would correspond, in a flat L–T ($\Lambda = 0$) interpretation, to

$$4.3 \frac{t_{B,r}(0)}{(9GM_0/2)^{1/3} t_p^{2/3}} + 3.75 \frac{t_{B,r}{}^2(0)}{(9GM_0/2)^{2/3} t_p^{4/3}}$$

$$+ 1.125 \frac{t_{B,rr}(0)}{(9GM_0/2)^{2/3} t_p^{1/3}} \approx -1 \pm 0.1. \tag{5.32}$$

Such a result would imply a negative value for at least one of the two quantities $t_{B,r}(0)$ or $t_{B,rr}(0)$, which would be an interesting constraint on the bang time function in the observer's neighbourhood. For instance, a function $t_B(r)$ decreasing near the observer would imply, for a source at a given $z < 1$, an elapsed time from the initial singularity that is longer in an L–T model than in the corresponding Friedmann one, i.e. an 'older' Universe. Therefore, a decreasing $t_B(r)$ has, in an L–T universe, an effect analogous to that of a positive cosmological constant in a Friedmann cosmology. They both make the observed Universe look 'older'.

5.1.4 Apparent acceleration

When examining the magnitude–redshift relation obtained from the SN Ia with no a priori idea about which model would best describe our Universe, a straight reading of the data does not exclude the possibility of challenging the cosmological principle, i.e. these data can fit other models than the FLRW ones with cosmological constant or dark energy.

As an example, it has been shown that any magnitude–redshift relation can be reproduced by L–T models with no cosmological constant, provided the parameter functions defining these models satisfy the constraints issuing from observations. However, such simple models have been given as an illustration and this does not preclude the more general conclusion stated above, i.e. that many other non-Friedmannian models might reproduce the effect of a dark energy component.

When reasoning in the framework of Friedmannian cosmology, the dimming of the supernovae is associated with an acceleration of the Universe expansion. This is why a number of authors have focussed on the issue of either demonstrating or ruling out an acceleration of the expansion rate in inhomogeneous models.

Some of them tried to derive (Flanagan, 2005; Hirata and Seljak, 2005) or rule out (Apostolopoulos *et al.*, 2006; Moffat, 2006) no-go theorems, stating that a locally defined expansion can never be accelerating in models where the cosmological fluid satisfies the strong energy condition.

However, when spatially averaged so as to reproduce a Friedmann-like behaviour, a physical quantity associated with the expansion rate behaves quite differently (Moffat, 2006a,b; Kai *et al.*, 2007). Now, Ishibashi and Wald (2006) argued that an averaged quantity representing the scale factor or the deceleration parameter may accelerate without there being any observable consequence. This issue was also studied by Räsänen (2006). Moreover, one can find realistic inhomogeneous models which fit the supernova observations and have positive values of volume deceleration (Enqvist and Mattsson, 2007; Bolejko and Andersson, 2008). Therefore, it is very difficult to deduce general rules from such theorems.

Another pitfall of this method was pointed out by Romano (2007a). He showed that an L–T model with a positive averaged acceleration can require averaging on scales beyond the event horizon of a central observer. In such cases, the averaging procedure does not preserve the causal structure of spacetime and can lead to the definition of locally unobservable average quantities. Sussman (2008) proved that the necessary condition to obtain volume acceleration in the L–T model is $E > 0$. When averaging on Mpc scales with inhomogeneities of realistic features, a large-amplitude function E which is required to obtain volume acceleration leads to a very large amplitude of the bang time function, making the age of the Universe unrealistically small (Bolejko and Andersson, 2008).

This reinforces the statement that a positive averaged acceleration of an underlying inhomogeneous Universe is in general not equivalent to a positive acceleration inferred from observations analysed within a Friedmannian scheme.

The notion of a deceleration parameter is not uniquely defined in an inhomogeneous scheme. Hirata and Seljak (2005) proposed four different definitions of such a parameter. Apostolopoulos *et al.* (2006), examining the effect of a mass located at the centre of a spherically symmetric configuration on the dynamics of a surrounding dust cosmological fluid, showed that, for an observer located away from the centre, (i) a central overdensity leads to acceleration along the radial direction and deceleration perpendicular to it, (ii) a central underdensity leads to deceleration along and perpendicular to the radial direction. This demonstrates that, even locally, the effect of inhomogeneities on the dynamics of the Universe is not trivial.

To understand intuitively how inhomogeneities can mimic an accelerated expansion, it is interesting to follow Tomita (2000, 2001, 2003) who considered a cosmological model composed of a low-density inner homogeneous region connected at some redshift to an outer homogeneous region of higher density. Both regions decelerate, but, since the void expands faster than the outer region, an apparent acceleration is experienced by the observer located inside the void.

We therefore conclude that the computation of some local quantity such as the deceleration parameter (possibly subsequently averaged) behaving the same way as in FLRW models with dark energy can lead to spurious results (Geshnizjani and Brandenberger, 2002; Räsänen, 2004; Bene *et al.*, 2006) and must therefore be avoided. Actually, what we observe is the dimming of the SN

Ia accounted for by the magnitude–redshift relation. 'Acceleration' is a mere consequence of the homogeneity assumption. This is the reason why we shall only consider, in the following, works aimed at studying this magnitude–redshift relation.

5.2 Direct and inverse problems

Several exact models have been proposed to mimic the supernova observations without dark energy: homogeneous void models (Tomita, 2001), Stephani models (Dąbrowski and Hendry, 1998; Barrett and Clarkson, 2000; Stelmach and Jakacka, 2001; Godłowski *et al.*, 2004) and the Locally Rotationally Symmetric inhomogeneous spacetime (Pascual-Sanchez, 1999). However, the most popular is the L–T model with which the most sophisticated studies have been performed up to now.

Two procedures for studying the cosmological constant problem with L–T models can be found in the literature:

- The direct way which consists in first guessing the form of the parameter functions defining a class of models supposed to represent our Universe with no cosmological constant, such as to write the dependence of these functions in terms of a limited number of constant parameters; then trying to fit these constant parameters to reproduce the observed SN Ia data or the luminosity distance–redshift relation of the standard ΛCDM model.
- The inverse problem where one considers the luminosity distance $D_L(z)$ as given by observations or by the ΛCDM model as an input and tries to select a specific L–T model with zero cosmological constant best fitting this relation.

5.3 The direct method

In this section, we examine some recent proposals using L–T classes of models and exemplifying the direct method. We consider only those aiming at reproducing the measured luminosity distance–magnitude relation (not 'acceleration') with zero cosmological constant and fulfilling the constraints which have been examined above:

- Theoretical constraints
  - No shell crossing (see Hellaby and Lake, 1985). These conditions hold for all models, with a central or off-centre observer.
  - In order to have a monotonically increasing $z(r)$, we need to have $\mathrm{d}z/\mathrm{d}r > 0$ and therefore, from (3.3), $R_{,tr} > 0$.
- Observational constraints
  - Constraints on the coefficients of the luminosity distance expansion from the SN Ia data, see (5.20)–(5.22).

- Constraints from the measured dependence of the Hubble parameter on red-shift.
- Constraints from the baryon acoustic oscillations analysed using the galaxy correlation function.
- Possibly, constraints issued from other cosmological data such as the CMB temperature fluctuation power spectrum, etc.

Since the problem is very complicated and even degenerate, the models which have been studied all pertain to special classes or subclasses of the L–T solutions, chosen a priori. The main goal is not to find THE inhomogeneous model for our local Universe, but to show that it is possible to reproduce the data with non-pathological models that are physically plausible and do not exhibit any cosmological constant.

5.3.1 Fitting the observations

In this subsection we confront three classes of inhomogeneous models that are free of dark energy, with observations of type Ia supernovae, with measurements of the baryon acoustic oscillations (BAO) and with the variation of the Hubble parameter with redshift (Bolejko, 2008; Bolejko and Wyithe, 2009).

Firstly, we consider observations of type Ia supernovae, which are taken from the Riess Gold Data Set (Riess *et al.*, 2007). However, the Riess supernova sample is scaled so that $H_0 = 65$ km s^{-1} Mpc^{-1} (also see the discussion in the appendix of Riess *et al.*, 2004), whereas in this subsection H_0 is chosen to be 70 km s^{-1} Mpc^{-1}. To avoid any consequent normalisation problems, we use the residual magnitude, $\Delta m = \mu - \mu_0 = 5 \log(D_L/D_{L0})$, where $\mu = 5 \log D_L + 25$ is the distance modulus, and μ_0 and D_{L0} are the expected values in an empty universe.

The second set of cosmological observations comprises measurement of the dilation scale of the BAO in the redshift space power spectrum. The dilation scale is defined as

$$D_V = \left[D_{A,0}{}^2 \frac{cz}{H(z)} \right]^2, \tag{5.33}$$

where $D_{A,0}$ is the comoving angular diameter distance and $H(z)$ is the Hubble parameter as a function of redshift. The Hubble parameter in (5.33), for the L–T model, is given by

$$H_R = R_{,rt}/R_r. \tag{5.34}$$

Based on the observation of 46 748 luminous red galaxies from the Sloan Digital Sky Survey, the dilation scale at $z = 0.35$ has been estimated to be 1370 ± 64 Mpc (Eisenstein *et al.*, 2005). It should be noted that this value was obtained within the framework of linear perturbations imposed on a homogeneous FLRW background. It is still an open question whether such an analysis is appropriate or should be carried out instead in the L–T background.[3] However, we provide the fit to this value to show that inhomogeneous models have no problems with fitting this measurement without the need for dark energy.

In addition to the geometric measurements described above, we use the constraints on the Hubble parameter presented by Simon, Verde and Jimenez (2005). These constraints are based on estimations of the age of the oldest stars in observed galaxies, as a function of redshift. If it is assumed that the age of the oldest star is comparable with the age of the Universe, then the Hubble constant can be derived from the derivative of the redshift with respect to time,

$$H(z) = -\frac{1}{(1+z)}\frac{\mathrm{d}z}{\mathrm{d}t};$$
(5.35)

as seen from (3.4) this quantity is equal to (5.34).

Given a particular model, the null geodesic equations can be solved to calculate the redshift, the luminosity distance D_L (fit to supernova data), the dilation distance D_V (fit to BAO), and the Hubble parameter H_R [to fit measurements of $\mathrm{d}z/\mathrm{d}t$, to make relation (5.35) more exact].

In this subsection we only focus on these three types of cosmological observations and do not consider the CMB constraints. This is because the last scattering surface is separated from regions where supernovae are observed by great distances (see the next subsection). A feature of inhomogeneous models is that they can provide quite different physical density distributions in each of the regions from which these two data sets are drawn. Thus, what is required to more strongly constrain inhomogeneous (and homogeneous) solutions to the apparent acceleration of the Universe is several sets of data that measure a range of observables at comparable redshifts. This is what the SN Ia, BAO and $H(z)$ offer – all these observational data are drawn from $z < 1.8$. The CMB power spectrum can then be fitted in two ways – either by modifying the model for higher redshifts (i.e. larger distances from the origin – see the next subsection) or as shown by Sarkar (2008) by assuming a more complex model of inflation. This last alternative requires that, at the distance of the last scattering surface, the expansion rate is of order $H(r = r_{\mathrm{CMB}}) = 46$ km s^{-1} Mpc^{-1}. As will be shown, all models which fit cosmological observations do indeed have decreasing expansion rates (as in Fig. 5.1).

Let us consider three classes of L–T models.

---

[3] The results of this section suggest the existence of a local Gpc scale inhomogeneity. Thus, if this inhomogeneity is real, then the analysis of the BAO (and other cosmological observations) should be carried out taking into account this inhomogeneity. If linear perturbations need to be considered, they should rather be imposed on the L–T background.

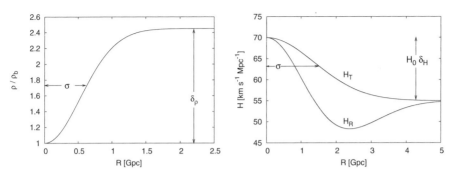

Fig. 5.1. The left panel presents the parametrisation of the density distribution within class I models ($t_B = 0$) at the current instant. The plotted profile corresponds to $\delta_\rho = 1.45$ and $\sigma = 0.75$. The right panel presents the parametrisation of the expansion rate (H_T) at the current instant within class II models, as well as the corresponding Hubble parameter H_R. The profiles shown correspond to $H_0 \delta_H = 15.02$ and $\sigma = 1.97$.

1. Class I: models with $t_B = 0$

Within these models the age of the Universe is everywhere the same. The density distribution at the current instant is assumed to be

$$\rho = \rho_b \left[1 + \delta_\rho - \delta_\rho \exp\left(-\frac{r^2}{\sigma^2} \right) \right], \tag{5.36}$$

where $\rho_b = 0.3 \times (3H_0^2)/(8\pi G)$. As can be seen, the density increases from ρ_b at the origin to $\rho = (1 + \delta_\rho)\rho_b$ at infinity. The left panel of Fig. 5.1 shows an example profile for which $\delta_\rho = 1.45$ and $\sigma = 0.75$. We note that the profile need not be extrapolated to infinity. Indeed, owing to the flexibility of inhomogeneous models the profile could be modified arbitrarily at larger distances in order to fit other types of observations, including those of the CMB.

Figures 5.2 and 5.3 show constraints on the parameters δ_ρ and σ based on the cosmological constraints described above. The confidence contours come from the χ^2 statistics. In each case the χ^2 for an L–T model with given parameters δ and σ has been calculated. Then, based on the χ^2 distribution with a given number of degrees of freedom, the probability P that the model is true in the light of the data has been calculated. Finally, the confidence contours, i.e. the confidence of ruling out the model $(1-P)$, has been found. These contours are presented in Figures 5.2 and 5.3. The upper-left, upper-right, and lower-left panels of Fig. 5.2 show individual constraints imposed by the observations of supernovae, of baryonic acoustic oscillations and of the Hubble parameter [i.e. from comparison of (5.34) with (5.35)]. The results of Fig. 5.2 show that large regions of parameter space are able to describe each of the three cosmological observations without including a cosmological constant. However, there are clear degeneracies between δ_ρ and σ in each case. The lower-right panel shows the joint constraints from all three cosmological observations. Although the best-fit model (the one with the lowest χ^2) is outside the 1σ contour, the

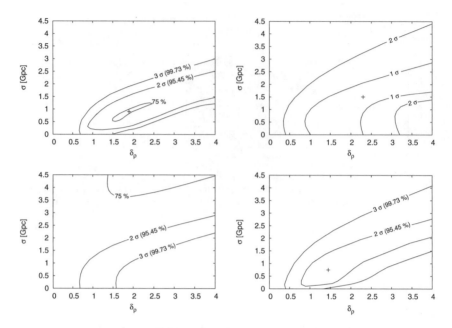

Fig. 5.2. Observational constraints on the parameters of Class I models, shown as contour plots of χ^2 for the model given the data. The confidence levels are set assuming a χ^2 distribution. Upper-left panel: constraints from supernova data (the cross denotes the χ^2 minimum, $\chi^2_{\min}/\text{dof} = 1.062$). Upper-right panel: constraints from SDSS LRG BAO data (the cross denotes parameters which provide a perfect fit to D_V). Lower-left panel: constraints from $H(z)$ measurements (the χ^2 minimum is outside the presented range). Lower-right panel: the combined constraints from SN, SDSS LRG BAO, and Hubble constant. At $\delta_\rho = 1.45$, $\sigma = 0.75$ (denoted by a cross) the confidence with which this model can be ruled out is 88.5%.

confidence of ruling out this model is not significantly high, and is equal to 88.5% (it is still within 2σ range). Within the parametrisation assumed, the best-fit void has a density contrast of $\delta_\rho \approx 1.5$ and a radius of $\sigma \approx 0.75$ Gpc (a profile corresponding to the example in the left panel of Fig. 5.1).

2. Class II: models with $t_B \neq 0$

   This family of models is defined by a given expansion rate at the current instant, of the form

   $$H_T = \frac{R_{,t}}{R} = H_0 \left[1 - \delta_H + \delta_H \exp\left(-\frac{r^2}{\sigma^2} \right) \right], \qquad (5.37)$$

   and by the density distribution of the form (5.36) with $\delta_\rho = \delta_H$. An example of such an expansion profile with $H_0\delta_H = 15.02$ km s^{-1} Mpc^{-1} and $\sigma = 1.97$ Gpc is presented in the right panel of Fig. 5.1, which also shows the corresponding H_R.

   The upper-left panel of Fig. 5.3 shows the constraints on the parameters from comparison with supernova data; the constraints from baryonic acoustic

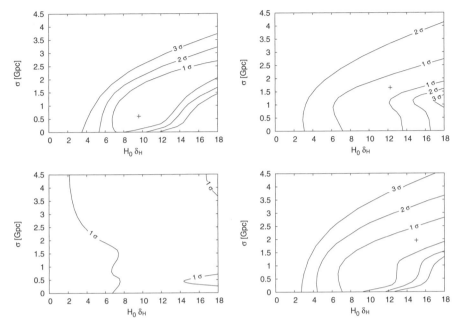

Fig. 5.3. Observational constraints on the parameters of Class II models, showing χ^2 contours and assuming a χ^2 distribution. Upper-left panel: constraints from SN data (the cross denotes the χ^2 minimum, $\chi^2_{\min}/\mathrm{dof} = 0.902$). Upper-right panel: constraints from SDSS LRG BAO (the cross denotes the parameters that provide a perfect fit to D_V). Lower-left panel: constraints from $H(z)$ measurements (the χ^2 minimum is outside the presented range). Lower-right panel: the combined constraints from SN, SDSS LRG BAO, $H(z)$. At $H_0\delta_H = 15.02$, $\sigma = 1.97$ (denoted by a cross) the confidence with which this model can be ruled out is 31.3% (for comparison the corresponding confidence of ruling out the ΛCDM model is 26.7%).

oscillation data are presented in the upper-right panel, and the constraints from comparing $H(z)$ in (5.35) with H_R are presented in the lower-left panel. The lower-right panel shows the joint constraints from all three cosmological observations. This family of models provides a better fit to available observations. The best-fit model has $H_0\delta_H \approx 15$ km s^{-1} Mpc^{-1} and $\sigma \approx 2$ Gpc. This model can be ruled out with a confidence of only 31.3%, which is very similar to that for the ΛCDM model (26.7%).

3. Class III: models with a homogeneous expansion rate at the current instant

In these models the expansion at the current instant is set to be homogeneous:

$$H_R(t_0, r) = \frac{R_{,rt}}{R_{,r}} = H_0 = 70 \text{ km s}^{-1} \text{ Mpc}^{-1}. \qquad (5.38)$$

The density and $t_B(r)$ are chosen to fit the supernova data. However, none of the attempts to obtain a satisfactory fit to the observational data succeeded. It seems that it is impossible to fit the supernova data with models where the expansion rate is currently constant. The best-fit model within the family of constant H models is the empty FLRW model (with $\Delta m = 0$).

5.3.2 Further constraints on inhomogeneous models from CMB data

In the standard approach, the CMB temperature fluctuations are analysed by solving the Boltzmann equation within linear perturbation around a homogeneous and isotropic FLRW model. In an inhomogeneous background one can do a similar analysis – employing the L–T model instead of an FLRW model. Alternatively, if it is assumed that the early Universe (before and up to the last scattering instant) is well described by an FLRW model, then the CMB power spectrum can be parametrised by (Doran and Lilley, 2002)

$$l_m = l_a(m - \phi_m),\tag{5.39}$$

where l_1, l_2, l_3 is a position of the first, second and third peak, and

$$l_a = \pi\frac{R}{r_s},\tag{5.40}$$

where R is the comoving distance to the last scattering surface and r_s is the size of the sound horizon at the last scattering instant. The sound horizon depends on Ω_b, Ω_m, Ω_γ and h. The size of the sound horizon can be calculated by the formula given in Eisenstein and Hu (1998). The function ϕ_m depends on Ω_b, Ω_m, h, n_s and the energy density of dark energy (n_s is the spectral index). The exact form of ϕ_m was given by Doran and Lilley (2002). The Ω_m and h are given by the ratio ρ/ρ_{cr} and $H/(100 \text{ km s}^{-1} \text{ Mpc}^{-1})$ at the last scattering surface (evaluated for the current instant).

Here we present a fit to the CMB data using the above formula. Let us consider four models,

- model A1:
  $\rho(t_0, r) = \rho_b \left[1 + 1.45 - 1.45 \exp\left(-\ell^2/0.75^2\right)\right]$,
  $t_B = 0$, $\Omega_b = 0.0445$, $n_s = 1$,
  where $\ell = r/\text{Gpc}$;
- model A1 + ring:
  $\rho(t_0, r) = \rho_b \left[1 + 1.45 - 1.45 \exp\left(-\ell^2/0.75^2\right)\right.$
  $\left. - 1.75 \exp\left\{-\left(\frac{\ell - 5.64}{0.926}\right)^2\right\}\right]$,
  $t_B = 0$, $\Omega_b = 0.08$, $n_s = 0.963$;
- model A2:
  $\rho(t_0, r) = \rho_b \left[3.33 + 1.4 \exp\left(-\ell^2/0.75^2\right)\right]$,
  $t_B = 0$, $\Omega_b = 0.0445$, $n_s = 1$;
- model A2 + ring:
  $\rho(t_0, r) = \rho_b \left[3.33 + 1.4 \exp\left(-\ell^2/0.75^2\right) + 0.08 \exp\left\{-\left(\frac{\ell - 7.5}{0.965}\right)^2\right\}\right]$,
  $t_B = 0$, $\Omega_b = 0.07$, $n_s = 0.963$.

Table 5.1. *The positions of the peaks (values of the multipole moment ℓ) of the CMB power spectrum.*

Model	First peak	Second peak	Third peak
A1	177.34	414.08	627.19
A2	216.59	503.37	754.28
A1 + ring	220.46	530.38	800.19
A2 + ring	220.43	530.51	779.76
WMAP	220.8 ± 0.7	530.9 ± 3.8	700–1000

Model A1 is the best-fit model of the Class I family considered in the section above (Fig. 5.2). The fit to the positions of the CMB peaks is presented in Table 5.1. As can be seen, model A1 does not fit the observed CMB power spectrum. However, we can modify the density distribution in such a way (by adding an underdense ring) that the CMB peaks can be reproduced – this is done in model A1 + ring. From these results we can see why the model of García-Bellido and Haugbølle (2008a), discussed below, has a larger radius for the local Gpc-scale void. In models where density increases up to some distance from the origin (local void type of models), a satisfactory fit to the first peak of the CMB power spectrum (i.e. distance to the last scattering instant) can be obtained if the mass of the Universe is decreased. This decrease can be obtained either by having a local void of a larger radius (as in García-Bellido and Haugbølle, 2008a, where $R \approx 2.5$ Gpc), or by having an additional underdense ring between the local void and the last scattering surface.

On the other hand, it is not only the mass that is important. To show this let us consider model A2. In it, the density decreases up to some distance from the origin. In this configuration we can also obtain a good fit to the position of the first peak if we increase the mass of the Universe.

The positions of other peaks strongly depend on Ω_b. To obtain a good fit to their positions in our model we need to increase the value of Ω_b beyond the value that is consistent with observation of light elements in the local Universe ($\Omega_b = 0.0445$; Cyburt, 2004) to $\Omega_b = 0.08$. However, due to the lower expansion rate at large distances, the physical baryon density $\Omega_b h^2$ remains close to the observed value, i.e. 0.0198 and 0.0212 for models A1 + ring and A2 + ring respectively.

The above results suggest that almost every model can be modified in such a way that a good fit to CMB data can be obtained. Models A1 and A2 are very different and yet after some modifications they are both able to fit the CMB data. Thus, the CMB data do not strongly constrain the properties of the local Gpc void. Each model of the local void considered in the previous section can be modified in such a way (by adding one or two rings between the local void and the last scattering surface) that the CMB power spectrum is recovered.

5.3.3 Alnes, Amarzguioui and Grøn's proposal

In a series of three papers (Alnes *et al.*, 2006, Alnes and Amarzguioui, 2006 and 2007), the authors studied a class of Universe models where the inhomogeneities take the form of a spherically symmetric underdense bubble, represented by a particular class of L–T models with $E(r) > 0$, which matches smoothly to a flat homogeneous Einstein–de Sitter background. In Alnes *et al.* (2006), the observer is assumed to be at the centre of the inhomogeneity and the goal is to reproduce both the SN Ia observations and the location of the first acoustic peak in the observed CMB temperature power spectrum. In the paper by Alnes and Amarzguioui (2006), the authors examine how far away from the centre the observer can be located to explain the alignment and magnitude of the lowest multipoles in the CMB map. This work is described in Sec. 7.3. In another paper by Alnes and Amarzguioui (2007), they investigate whether such an off-centre position of the observer might yield a better fit of the model to the SN Ia data than in Alnes *et al.* (2006). We report the results of this work in the present section.

First we summarise the presentation of the model as it is given in Alnes *et al.* (2006). The Einstein equations for the L–T solutions can be put in the form

$$\kappa\rho = H_\perp^2 + 2H_r H_\perp - \frac{\beta}{R^2} - \frac{\beta_{,r}}{RR_{,r}}, \tag{5.41}$$

$$-\kappa\rho = -6H_\perp^2 q_\perp + 2H_\perp^2 - 2\frac{\beta}{R^2} - 2H_r H_\perp + \frac{\beta_{,r}}{RR_{,r}}, \tag{5.42}$$

where $H_\perp \equiv R_{,t}/R$, $H_r \equiv R_{,tr}/R_{,r}$ and $q_\perp \equiv -RR_{,tt}/R_{,t}^2$.

The L–T region of the model is described by two functions renamed $\alpha(r) \equiv 2GM(r)$ and $\beta(r) \equiv 2E(r)$. Since they are interested in the late-time behaviour of the model, the authors define $t = 0$ as the time when the photons decoupled from the matter, i.e. the time of last scattering. Furthermore, they define $R(t = 0, r) \equiv R_0(r)$ and introduce a 'conformal time'[4] η by $\beta^{1/2}\mathrm{d}t = R\mathrm{d}\eta$. To represent an underdensity, i.e. a L–T model with negative spatial curvature, the authors integrate the Einstein equations for $\beta > 0$ which yields

$$R = \frac{\alpha}{2\beta}(\cosh\eta - 1) + R_0\left[\cosh\eta + \sqrt{\frac{\alpha + \beta R_0}{\beta R_0}}\sinh\eta\right], \tag{5.43}$$

$$\sqrt{\beta}t = \frac{\alpha}{2\beta}(\sinh\eta - \eta) + R_0\left[\sinh\eta + \sqrt{\frac{\alpha + \beta R_0}{\beta R_0}}(\cosh\eta - 1)\right]. \tag{5.44}$$

---

[4] The 'conformal time' η so defined is a function of both t and r, and in fact its definition should be written as $\partial\eta/\partial t = \sqrt{\beta}/R$. The term 'conformal time' suggests that it is a time coordinate, but in truth this is a parameter defined separately for each constant-r worldline. In the paper, though, the authors use it in a correct way.

The α and β functions are chosen so that they correspond to a smooth inter-polation between two homogeneous regions where the inner region has a lower matter density than the outer region, thus describing a spherical bubble in an otherwise homogeneous universe. They are written as

$$\alpha(r) = H_0^2 r^3 \left[\alpha_0 - \Delta\alpha \left(\frac{1}{2} - \frac{1}{2} \tanh \frac{r - r_0}{2\Delta r} \right) \right], \qquad (5.45)$$

$$\beta(r) = H_0^2 r^2 \left[\beta_0 - \Delta\beta \left(\frac{1}{2} - \frac{1}{2} \tanh \frac{r - r_0}{2\Delta r} \right) \right], \qquad (5.46)$$

where H_0 is the value of the Hubble parameter in the outer homogeneous region today, while α_0 and β_0 are the relative densities of matter and curvature in this region. Furthermore, $\Delta\alpha$ and $\Delta\beta$ represent the differences for α and β between the central and homogeneous regions, and r_0 and Δr specify the position and the width of the transition zone.

Since the function $R_0(r)$ can be chosen at will using the freedom of coordinate choice in L–T solutions, the authors choose $R_0 = a_* r$, where a_* is the scale factor of the FLRW model at recombination, in order to match the L–T solution to the homogeneous one at that epoch.

To reduce the number of degrees of freedom of the model and make the problem tractable, the authors restrict themselves to a very simple toy model: an under-dense spherically symmetrical region surrounded by a flat, matter-dominated universe. To implement this, they choose $\alpha_0 = 1$ and $\beta_0 = 0$. Moreover, they put $\Delta\alpha = -\Delta\beta$. This leaves four parameters, $\Delta\alpha$, r_0, Δr and the Hubble parameter at the origin, $H_r(0, t_0) = 100h$ km s^{-1} Mpc^{-1}, to be fitted to the observations.

To relate the α and β functions to observable quantities, they define relative matter and curvature densities from the generalised Friedmann equation (5.41) as

$$\Omega_m = \frac{\kappa\rho}{H_\perp^2 + 2H_r H_\perp}, \qquad \Omega_k = 1 - \Omega_m. \qquad (5.47)$$

Central observer

In the paper by Alnes *et al.* (2006), the observer is located at the centre of the underdensity. We give below a short overview of this analysis since it takes into account some interesting cosmological features seldom found in such stud-ies and since its results are utilised in the off-centre observer case (Alnes and Amarzguioui, 2006 and 2007) which will be described further on.

Equations (3.3) and (3.4), which define the photon path, can be recast into Eqs. (19) and (20) of Alnes *et al.* (2006). Then, to determine the luminosity distance–redshift relation to be compared with the SN Ia observations, these equations are numerically integrated, with the initial conditions: $T(0) = t_0 \equiv t(r = 0)$ and $z(0) = 0$. The location of the last scattering surface, i.e. the position of the CMB photons we observe today at the time of last scattering, is given by

$T(r_*) = 0$ and t_0 is defined by $z(r_*) = z_* \simeq 1100$. The luminosity distance follows from (5.17) and the angular diameter distance is

$$D_A(z) = R(t(z), r(z)). \tag{5.48}$$

Now, to confront their model with CMB observations, the authors would, in principle, need to study perturbations in an inhomogeneous universe. However, since this model is homogeneous outside a limited region at the centre, they assume that the evolution of the perturbations is identical to that in a homogeneous universe until the time of recombination. This allows them to use the standard results for the scale of the acoustic oscillations on the last scattering surface. Anyway, the angular diameter distance, which converts this scale into an observed angle on the sky, is sensitive to the inhomogeneity at the centre.

These authors also calculate the position of the first Doppler peak of the power spectrum in their model, and show that agreement with observations is not hard to produce.

However, there are a lot of possible choices for the parameters of the model that give a very good fit to both the supernova data and the position of the first acoustic peak in the CMB temperature power spectrum. Remember we have shown in Sec. 5.1 that the problem consisting in deriving an L–T model from the SN Ia data is degenerate. Therefore, it is not very surprising that the addition of such a slight constraint from the CMB spectrum is insufficient to remove this degeneracy. For this reason the authors add another constraint: that their model be able to reproduce the mass density parameter at the origin, Ω_{m0}, measured from observations of galaxies with redshifts $z < 0.12$ by the 2dF team. They propose, as their 'standard model', a solution which gives an underdensity at the centre, $\Omega_{m0} = 0.2$, compatible with the results of the 2dF survey, $\Omega_{m0} = 0.24 \pm 0.05$.

The authors also try to use the scale of the baryon oscillations detected in the SDSS galaxy power spectrum (Eisenstein *et al.*, 2005) to constrain their model even further. The physical length scale associated with these oscillations is fixed by the sound horizon at recombination. Therefore, measuring how large this length scale appears at some redshift in the galaxy power spectrum might allow one to constrain the time evolution of the universe from recombination to the time corresponding to this redshift. Anyway, it is unclear how to use the values calculated by Einsenstein *et al.* (2005), since these authors derived their results in the framework of a ΛCDM model. As we already stressed in a previous footnote, ideally, one would need to reproduce the analysis of the SDSS data assuming an inhomogeneous model.

The results of the analysis of such a class of models with a central observer are that the underdensity might extend about 1.35 Gpc outwards. A very good fit to the supernova data is obtained if the transverse Hubble parameter $H_\perp$ is allowed to decrease with distance from the observer. This fit is even better than for the ΛCDM model. On the other hand, a good fit to the location of the

Table 5.2. *Parameters and features of the best fit model of Alnes et al. (2006) with a central observer.*

Parameter or feature	Symbol	Value
Density contrast	$\Delta\alpha$	0.90
Transition point	r_0	1.35 Gpc
Transition width	$\Delta r/r_0$	0.40
Fit to supernova data	χ^2_{SN}	176.5
Position of the first CMB peak	$\mathcal{S}$	1.006
Age of the Universe	t_0	12.8 Gyr
Relative density inside underdensity	$\Omega_{m,\text{in}}$	0.20
Relative density outside underdensity	$\Omega_{m,\text{out}}$	1.00
Hubble parameter inside underdensity	h_{in}	0.65
Hubble parameter outside underdensity	h_{out}	0.51
Physical distance to the last scattering	D_{LSS}	11.3 Gpc

first peak of the CMB power spectrum is reached if the Universe is assumed to be flat with a value 0.51 for the Hubble parameter outside the inhomogeneity. Therefore, interpolating between these two limiting behaviours might provide a good model to account both for the SN Ia observations and the location of the first CMB peak. The values of the parameters for the best-fit model are given in Table 5.2.

Off-centre observer

Alnes and Amarzguioui (2006 and 2007) consider the best-fit model of Alnes *et al.* (2006), but now they put the observer away from the centre.

In their first paper they investigate the influence of such an observer location on the CMB temperature power spectrum (we will come back to this in Chapter 7), and in the second paper they try to improve the fit of their model to the SN Ia data by moving the observer off centre, and also to see how much these data constrain such a location.

By moving the observer off the centre, one adds three new degrees of freedom as regards the model with a central observer, the distance from the centre, r_0, and the two angles which specify the displacement direction. Since the explicit angular dependence of the observed supernovae is available in the data, the most appropriate way to do the analysis is to minimise the χ^2 values with respect to the two angles for each r_0.

The consequence of such an observer location is that the distance measures become anisotropic. The explicit effect on the expression for the angular diameter distance was analysed by Ellis *et al.* (1985) and Humphreys *et al.* (1997). The expression, as given in the paper by Humphreys *et al.* (1997), reads

$$D_A{}^4 \sin^2\gamma = \tilde{g}_{\gamma\gamma}\tilde{g}_{\xi\xi} - \tilde{g}_{\gamma\xi}^2, \tag{5.49}$$

where the $\widetilde{g}_{\mu\nu}$ are the metric coefficients in the observer rest frame and γ and ξ correspond to the polar and azimuthal angles in this frame. Such a frame can be constructed using the light cones specified in Sec. 3.1.3.

After integrating the geodesic equations of Sec. 3.1.3, which determine the trajectories of the infalling photons toward the off-centre observer, one proceeds with the diameter distance calculation. The metric $\widetilde{g}_{\mu\nu}$ in the observer's local coordinates is obtained by the coordinate transformation

$$\widetilde{g}_{\mu\nu} = g_{\rho\sigma} \frac{\partial x^\rho}{\partial \widetilde{x}^\mu} \frac{\partial x^\sigma}{\partial \widetilde{x}^\nu}, \tag{5.50}$$

which yields the following components,

$$\widetilde{g}_{\gamma\gamma} = g_{rr} \left(\frac{\partial \hat{r}}{\partial \gamma} \right)^2 + g_{\vartheta\vartheta} \left(\frac{\partial \hat{\vartheta}}{\partial \gamma} \right)^2, \tag{5.51}$$

$$\widetilde{g}_{\xi\xi} = g_{\varphi\varphi} = R^2 \sin^2 \vartheta, \tag{5.52}$$

$$\widetilde{g}_{\gamma\xi} = 0. \tag{5.53}$$

Substituting these equations into (5.49) yields the expression for the angular diameter distance which reads

$$D_A{}^4 = \frac{R^4 \sin^2 \vartheta}{\sin^2 \gamma} \left[\frac{R_{,r}{}^2}{R^2 (1 + 2E)} \left(\frac{\partial \hat{r}}{\partial \gamma} \right)^2 + \left(\frac{\partial \hat{\vartheta}}{\partial \gamma} \right)^2 \right]. \tag{5.54}$$

By the reciprocity theorem (5.5) it is now straightforward to obtain the luminosity distance D_L.

The results are compared to the data points and error bars from the Riess et al. (2004) Gold Set. For the best-fit model of Alnes et al. (2006), it is found that the minimised χ^2 value is smallest for an observer located at a physical distance around 94 Mpc from the centre of the inhomogeneity in the direction $(l, b) = (271, 21)$ in galactic coordinates. However, the improvement turns out to be small compared to the fit with the central observer model (the minimal χ^2 is only reduced from 176.2 to 174.9). Therefore, the current data do not offer any substantial evidence for an off-centre observer in this model.

Moreover, the authors find that the χ^2 is lower for off-centre observers with radial distances out to about 225 Mpc. From this they conclude that anisotropies in the SN Ia data do not strongly constrain the observer's distance from the centre. They suggest that this is partly due to the fact that there are too few supernovae in the studied sample, hoping that in the future larger and better samples would become available to allow them to draw stronger conclusions regarding the anisotropy of the local Universe.

5.3.4 GBH model

In García-Bellido and Haugbølle (2008a) a new class of L–T models (the GBH models) is considered. It is apparently similar to the one studied by Alnes *et al.* (2006), however differing in the details. These models, describing a local void in an otherwise spatially asymptotically flat universe, are completely specified by their matter content $\Omega_m(r)$ and their rate of expansion $H(r)$,[5]

$$\Omega_m(r) = \Omega_{\text{out}} + \left(\Omega_{\text{in}} - \Omega_{\text{out}}\right)\left(\frac{1 - \tanh[(r - r_0)/2\Delta r]}{1 + \tanh[r_0/2\Delta r]}\right), \qquad (5.55)$$

$$H(r) = H_{\text{out}} + \left(H_{\text{in}} - H_{\text{out}}\right)\left(\frac{1 - \tanh[(r - r_0)/2\Delta r]}{1 + \tanh[r_0/2\Delta r]}\right). \qquad (5.56)$$

It is governed by 6 parameters:

Ω_{out}	determined by asymptotic spatial flatness,
Ω_{in}	determined by LSS observations,
H_{out}	determined by CMB observations,
H_{in}	determined by HST observations,
r_0	characterises the size of the void,
Δr	characterises the transition to uniformity.

One parameter is fixed, $\Omega_{\text{out}} = 1$, the other five are left to vary freely in the parameter scans.

These authors consider also a more constrained model, in which the Big Bang is simultaneous. This can be attained simply (see the original article for details) by a choice of $H(r)$,

$$H(r) = H_0\left[\frac{1}{\Omega_K(r)} - \frac{\Omega_m(r)}{\sqrt{\Omega_K^3(r)}}\sinh^{-1}\sqrt{\frac{\Omega_K(r)}{\Omega_m(r)}}\right], \qquad (5.57)$$

so that $t_B = c\,H_0^{-1}$ is universal, for all observers, irrespective of their spatial location. Note that in this case one has less freedom than in the previous one, since now there is only one arbitrary function, $\Omega_m(r)$, and there is one free parameter less.

The goal is to confront this model with the largest series of currently available observations: supernova data, cosmic microwave background, baryon acoustic oscillations; also requiring it to obey three additional priors: the observed lower age limit on globular clusters in the Milky Way (11.2 Gy), the HST key project measure of the local value of the Hubble parameter ($H_{\text{in}} = 72 \pm 8$) and the gas fraction as observed in galaxy clusters. Here, the observer is located at the centre of the model.

---

[5] $H_0(r)$ in the authors' notation.

They propose that the Hubble parameter $H(z)$ is one of the main observables that will in future help to decide between the different possible scenarios (modifications of gravity, extra energy component, cosmological constant, or nonhomogeneous cosmological model). Assuming the background is a flat FLRW cosmology, they write it as

$$\frac{H_{T,L}{}^2(z)}{H_{\rm in}{}^2} = (1+z)^3 \Omega_{\rm in}$$

$$+ (1 - \Omega_{\rm in}) \exp\left[3 \int_1^{1+z} {\rm d}\log(1+z')(1 + w_{\rm eff}^{T,L}(z'))\right], \qquad (5.58)$$

where $H_{\rm in}$ and $\Omega_{\rm in}$ are the expansion rate and matter density as observed at $z = 0$, the labels T and L denote the transverse and longitudinal expansion rates, defined as $H_T = R_{,t}/R$ and $H_L = R_{,tr}/R_{,r}$, and the $w_{\rm eff}^{T,L}$ are defined in terms of the derivative of (5.58),

$$w_{\rm eff}^{T,L}(z) = -1 + \frac{1}{3}\frac{{\rm d}\log\left[H_{T,L}{}^2(z)/H_{\rm in}{}^2 - (1+z)^3\Omega_{\rm in}\right]}{{\rm d}\log[1+z]}, \qquad (5.59)$$

where it has been assumed that $H_{T,L}{}^2(z)/H_{\rm in}{}^2 - (1+z)^3\Omega_{\rm in} > 0$. The authors point out that $w_{\rm eff}^{T,L}(z)$ is the equation of state parameter in the dark energy interpretation, while in the L–T or modified gravity interpretations it is an empirical observational signature that expresses the difference between the measured expansion rate and the expansion rate we ascribe to the observed matter.

Another key outcome of this work is that the baryon acoustic oscillation signal depends partly on H_L, while supernova observations are only related to H_T through its dependence on $D_L = (1+z)^2 D_A = (1+z)^2 \exp(\int H_T {\rm d}t)$. Interestingly, the variation and derivative of $w_{\rm eff}^{T,L}(z)$ is quite large in the best-fit L–T models, showing that a precise low-redshift supernova survey sensitive to H_T or a fine-grained BAO survey sensitive to H_L could rule out or reinforce the model in the near future. Conversely, if a disagreement between w as observed by supernovae and w as observed through the BAOs is found, this could be a hint of inhomogeneous expansion rates.

One can directly compute $w_{\rm eff}^{T,L}(z)$ in the limiting cases $z = 0$ and $z \gg 1$ for asymptotically flat L–T models:

$$w_{\rm eff}^{T,L}(z) = \begin{cases} -\dfrac{1}{3} + \dfrac{2}{3}\dfrac{cH_{,r}(0)}{(1-\Omega_{\rm in})H_{\rm in}^2} & \text{for } z = 0, \text{ and } H = H_T \\[2ex] -\dfrac{1}{3} + \dfrac{4}{3}\dfrac{cH_{,r}(0)}{(1-\Omega_{\rm in})H_{\rm in}^2} & \text{for } z = 0, \text{ and } H = H_L \\[2ex] 0 & \text{for } z \gg 1 \end{cases}, \qquad (5.60)$$

where the asymptotic convergence of the L–T metric to a Friedmann metric giving ${\rm d}z = -{\rm d}a/a$ has been used (here a denotes the scale factor of the Friedmann background). It appears that to have $w \ll -1/3$ at low z requires either a significant negative gradient in $H(r)$, or $\Omega_{in} \sim 1$.

In a subsequent work, Garcia-Bellido and Haugbølle (2008b) have studied the kinematic Sunyaev–Zel'dovich (kSZ) effect observed for nine distant galaxy clusters. Assuming we are located at the centre of a constrained GBH void, the large CMB dipole which should be observed in the reference frame of such off-centre clusters would actually manifest itself for us as a kSZ effect. The authors show that the kSZ observations give far stronger constraints on their model than other observational probes and infer that the void permitted by the existing data cannot be larger than ~ 1.5 Gpc. They stress that kSZ surveys to be completed in the near future will put still more stringent constraints on the size of the void in their model. On the other hand, if this class of models were to appear to be compatible with observations, future kSZ surveys could give a unique possibility of directly reconstructing the expansion rate and the underdensity profile of the void.

The general conclusion of their work is that the current observations do not exclude the possibility that we live close to the centre of a large void with size of order of a few Gpc, with matter density well below the average, and a local expansion rate of 71 km s^{-1} Mpc^{-1}.

5.3.5 L–T Swiss-cheese models

The first Swiss-cheese model which can be found in the literature was designed by Einstein and Straus (1945) to study the gravitational fields surrounding stars. The vacuoles, cut out from the Friedmann background, were modelled by a static Schwarzschild solution. Such models were subsequently used to study, among other effects, the influence of inhomogeneities on the magnitude–redshift relation (see e.g. Kantowski, 1969, and a series of papers by Nottale, 1982, 1983). However, since Schwarzschild's is a static solution, any influence of the vacuole expansion remained unseen in such models, and the magnitude of the reported effects was very low (as an example, Nottale, 1982b, using a very simplified vacuole model, found an observable amplification by medium-density clusters or by superclusters of galaxies of only a tenth of a magnitude).

The recent appearance in the literature of Universe models constructed as L–T Swiss-cheese models, where the inhomogeneities are represented by L–T spherical regions within a homogeneous background where the matter is assumed continuously distributed with densities both below and above the average, allows us to account for this vacuole expansion. These local L–T bubbles exactly match to the Friedmann background. The main shortcoming of such models, as regards the issue of solving the cosmological constant problem, is that they can produce surprising cancellations. It is well known that when the two metrics, L–T and Friedmann, are properly matched together, the metric outside the patch is insensitive to the details of what happens inside. This feature suppresses any backreaction on the homogeneous region which still evolves exactly as a Friedmann universe. For this reason, such models exhibit only corrections due to light propagation inside the patches, usually called the Rees–Sciama effect.

Note however that, at variance with what is claimed in Biswas and Notari (2008), such a drawback is not due to spherical symmetry but to the fact that the inhomogeneous patches are exactly matched to the background. We have shown indeed in Chapter 4 that within such Swiss-cheese models in which the inserted inhomogeneous patches were not L–T bubbles but parts of the Szekeres spacetime, with no symmetry, there was still no influence on the outside region. Moreover, one can have a sort of backreaction with both spherical and non-spherical inhomogeneity, provided an ordinary matching with the metric and second form being continuous is not required, but only the continuity of the metric. Then the boundary of the inhomogeneous island does not comove with the Friedmann background, giving rise to interesting phenomena. This was first found by Sato and coworkers [see e.g. Sato (1984) for a review]. Such an effect is also present in the models described in Chapter 4.

The issue of the influence of the spherical symmetry of the holes in L–T Swiss-cheese models has been addressed in the papers by Marra *et al.* (2007, 2008), where such a class of models was studied. A summary of these works is given in the following.

The first authors, to our knowledge, to have considered L–T Swiss-cheese models to deal with the 'cosmological constant problem' were Kai *et al.* (2007). However, their aim was to reproduce an accelerated cosmic expansion and not the luminosity distance–redshift relation. Therefore, the constraints they found on their model cannot be considered as relevant to actually reproducing the observations.

Brouzakis et al.'s model

Brouzakis *et al.* (2007, 2008) studied a model where the L–T patches consist of collapsing or expanding regions matched to a Friedmann background. Since they intended that the observer and the sources do not occupy special positions, they placed both within the Friedmann domain. Therefore, light is emitted and received within homogeneous regions, while it crosses several inhomogeneous bubbles, with size of order $10h^{-1}$ Mpc or more, along its path.

In Brouzakis *et al.* (2007), the studied L–T models were of the constant bang time type, i.e. $t_B(r) = 0$. Central overdense regions become denser with time, with underdense spherical shells surrounding them. Central underdense regions turn into voids, surrounded by massive shells. The authors recalled the derivation of the optical equations, as given by Sachs (1961), which determine the evolution of the beam expansion θ and shear σ as light propagates and derived their specific form for L–T models, which reads (in our notation)

$$\frac{d\theta}{d\lambda} = -4\pi G\rho(k^t)^2 - \theta^2 - \sigma^2, \tag{5.61}$$

$$\frac{d\sigma}{d\lambda} + 2\theta\sigma = 4\pi G(k^\varphi)^2 R^2 \left(\rho - \frac{3M(r)}{4\pi R^3}\right), \tag{5.62}$$

while the equation for the beam area A can be written as

$$\frac{1}{\sqrt{A}}\frac{d^2\sqrt{A}}{d\lambda^2} = -4\pi G\rho(k^t)^2 - \sigma^2. \tag{5.63}$$

They derived a simple model of spherical collapse for the cases of overdense and underdense L–T regions, which gave the evolution of both density profiles. Then, the influence of light propagation toward the centre of such collapsing regions upon the luminosity distance was evaluated by numerically integrating the geodesic and optical equations. The results are that the presence of underdense central regions, consistent with the observation of large, approximately spherical voids in the matter distribution of the Universe, is expected to lead to an increase of the luminosity distance relative to the homogeneous case. The opposite is expected if the central regions are overdense.

It is therefore reasonable to suggest that the dominance of voids, as deduced from observations, can yield increased luminosity distances. The question is whether this can induce an effect which might explain the observed luminosity curves for supernovae. However, a description of the Universe as being composed of large voids surrounded by matter concentrated in thin shells implies, in this L–T Swiss-cheese model, a relative increase in the luminosity distance at redshifts of order unity of only a few %. The deviation of the luminosity distance from its value in a homogeneous universe was estimated in the extreme case where light travels through the centres of the encountered inhomogeneities. An absolute upper bound upon this increase can be derived from (5.63). The focussing of a beam is minimised if the shear is negligible and the energy density set to zero in this equation. In this L–T Swiss-cheese model, this idealised situation can be achieved if the central underdense regions of the inhomogeneities become totally empty after a long evolution, while overdense thin shells develop around them. To derive the upper bound, one can set the energy density arbitrarily to zero in the optical equation only, since it still drives the cosmological expansion through (2.2) (with a zero Λ). Neglecting the shear, the luminosity distance as a function of redshift can be obtained analytically, assuming that the background expansion is given by the standard Friedmann equation involving the average density. When comparing it to the luminosity distance in the corresponding Friedmann model, one finds

$$\frac{D_L}{D_{L,F}} = \frac{1}{5} \frac{(1+z)^2 - (1+z)^{-3/2}}{1+z - (1+z)^{1/2}}. \tag{5.64}$$

For redshifts near 1, this expression gives a maximum increase in the luminosity distance of around 24% (we are not sure why the authors give it as 10%) to be compared to the measured increase which is around 30%. The open question is whether an alternative model of large-scale structures could result in a larger effect. This is discussed in the conclusions of Brouzakis *et al.* (2007).

In Brouzakis *et al.* (2008), the matter distribution, even though inhomogeneous, is more evenly distributed than in the above case. In particular, each spherical region has a central underdensity surrounded by an overdense shell. The densities in the voids and the background are comparable at early times and differ by a factor of order 1 during the later stages of evolution. The beam shear is negligible in the calculations, and the main effect arises from the variations of the beam expansion due to the inhomogeneities. Also, more general beam geometries are considered, in which the light paths have random impact parameters relative to the centres. In this work, a more detailed statistical analysis is performed and, for a given redshift, the width of the distribution of luminosity distances is evaluated.

The form of this distribution is quantified in terms of two parameters: the width δ_d and the location of the maximum $\delta_m > 0$. The first one characterises the error induced to cosmological parameters derived in the standard way, while the second one determines the bias in such determinations. However, the values of δ_d and δ_m are very small for the considered inhomogeneity scales. Furthermore, the shift in the average value δ_m is always smaller than the standard deviation δ_d. Both increase with the inhomogeneity scale and the redshift. But, even for inhomogeneities with large length scales, comparable to the horizon scale, and redshift $z = 2$, the shift and distribution width are only of the order of a few %. They can be compared to those generated by gravitational lensing at scales typical of galaxies or galaxy clusters in a standard Swiss-cheese model where the mass of each inhomogeneity is concentrated in a very dense object located at its centre and where the inhomogeneous patch is modelled by a Schwarzschild metric. Here, the typical values of δ_d and δ_m are larger by at least an order of magnitude than in this L–T Swiss-cheese model. The reason is that, in this last model, the density contrast is only of order 1.

The conclusion is that, within this L–T Swiss-cheese model, the presence of large length scale inhomogeneities and density contrast of order 1 does not influence significantly the propagation of light if the source and the observer possess random locations.

In a subsequent article, Brouzakis and Tetradis (2008) provided an analytical estimate of the effect of an L–T inhomogeneity on light beams. The relative deviations of travelling time, redshift, beam area and luminosity distance from their values in a homogeneous model depend on the ratio $\overline{H}$ of the inhomogeneity radius to the horizon distance. For an observer located at the centre of the inhomogeneous patch, the deviations are of order $\overline{H}^2$. For an outside observer, the deviations of crossing time and redshift are of order $\overline{H}^3$, while the deviations of the beam area and of the luminosity distance are of order $\overline{H}^2$. Therefore, in both cases, the deviation of the luminosity distance is of order $\overline{H}^2$. This deviation experiences an increase, respectively a decrease, while crossing an underdensity, respectively overdensity.

However, a cancellation due to positive and negative contributions during multiple crossings as advocated by the authors is not obvious since we know that voids are observed to occupy a larger volume than condensed structures and are suspected to evolve more rapidly (see Chapter 4). Therefore, the conclusion of these articles must be considered with care.

Biswas and Notari's model

An analogous L–T Swiss-cheese model has been studied by Biswas and Notari (2008). It consists of blobs of the void type, where the L–T metric satisfies $E(r) > 0$ and $t_B(r) = 0$, cut out from a flat Einstein–de Sitter universe. The photon propagation is studied in the cases when the curvature of the patches is either small or large. In both cases the effects are very small if the observer sits outside the void, strengthening therefore Brouzakis *et al.*'s results, while they are large if she sits inside. (This is a property of Swiss-cheese models. Effects are always larger inside the inhomogeneous patch than outside it. See also Sec. 7.1, where it is shown that the CMB temperature fluctuations are always larger when they are measured inside the perturbed region than outside it.)

Marra et al.'s model

Another Swiss-cheese model, where L–T bubbles with radius 350 Mpc (roughly 25 times smaller than the radius of the visible Universe) are also inserted into an Einstein–de Sitter background, was studied by Marra *et al.* (2007). In their main model the bubbles are adjoining and laid out on a cubic lattice, as sketched in Fig. 3 of Marra *et al.* (2007).

The initial conditions for each sphere are specified for every shell at the same moment of time $\bar{t}$. The initial density, $\rho(r, \bar{t})$, is chosen to be

$$\rho(r, \bar{t}) = A \exp[-(r - r_M)^2/2\sigma^2] + \epsilon \quad (r < r_h),$$
$$\rho(r, \bar{t}) = \rho_{ES}(\bar{t}) \quad (r > r_h) \tag{5.65}$$

where $\epsilon = 0.0025$, $r_h = 0.042$, $\sigma = r_h/10$, $r_M = 0.037$, $A = 50.59$ and $\rho_{ES}(\bar{t}) = 25$. It exhibits a low-density interior, surrounded by a Gaussian density peak near the boundary, that matches smoothly to the exterior Friedmann density, and such that the matter density ϵ in the centre is roughly 10^4 times smaller than in the Friedmann cheese. To have a realistic evolution, it is also demanded that there are no initial peculiar velocities. This fixes the arbitrary curvature function $E(r)$ to a positive value small compared to unity.

Evolving this model from $t = \bar{t} = -0.8$ to $t = 0$ (with the 'Big Bang' time $t_B = -1$ and $t = 0$ being today), one can see that the inner almost empty region expands faster than the outer (cheese) one. The density ratio between the cheese and the interior of the hole increases by a factor of 2. The evolution is realistic. Matter is falling toward the peaks in density. Overdense regions start contracting and become thin shells, mimicking structures, while underdense regions become larger, mimicking voids, and eventually they occupy most of the volume. Because of the distribution of matter, the inner part of each hole is expanding faster than the cheese and the interpolating overdense region is squeezed by it. A shell crossing eventually happens when shells are so squeezed that they occupy the same physical position, i.e. when $R_{,r} = 0$ (this is consistent with what we found for the models of void evolution studied in Sec. 4.4.3). However, the authors assert that nothing happens to the photon other than passing through more shells at the same time.

Because the aim of this work is to calculate the luminosity distance–redshift relation $D_L(z)$ in order to understand the effects of inhomogeneities on observables, the next step is to study the propagation of photons in this model. Three cases are considered: the observer is just outside the last hole, in the Friedmann cheese, looking at photons passing through the holes; the observer is in a hole on a high-density shell; the observer is in the centre of a hole.

Since the photons subtend an angle α at the observer after passing through the centres of all the holes, the equations describing the photon paths are analogous to those established in Sec. 3.1.2, with the value of the affine parameter λ set to zero at the border of the last hole. Given an angle α, the equations can be solved, matching together the solutions between one hole and the other, which generates the solution in the form $t(\lambda)$, $r(\lambda)$, $\phi(\lambda)$ and $z(\lambda)$, from which the observables of interest can be calculated.

The difference for the photon paths between the cases 'observer on the border' and 'observer in the hole' is that in the second case the observer has a peculiar velocity with respect to a Friedmann observer passing by the same point, which is not the case for the observer on the border who is comoving with the cheese. This makes the inside observer see an anisotropic CMB. However, this effect is taken into account in the solution to be found. There remains the effect of light aberration which changes the angle α seen by the inside observer with respect to the angle α_{ES} seen by a Friedmann observer. The relation between α and α_{ES} is given by the relativistic aberration formula:

$$\cos \alpha_{ES} = \frac{\cos \alpha + v/c}{1 + v/c \, \cos \alpha}. \tag{5.66}$$

The angular diameter and luminosity distances follow.

The results are given for different cases and compared with respect to the EdS universe and the ΛCDM one. For the observer on the border, one model exhibits five holes and the other only one big hole five times bigger in size than each of the previous holes. For the observer in the hole, only the five-hole case is considered. The observables studied are the redshift $z(\lambda)$, the angular diameter distance $D_A(z)$, the luminosity distance $D_L(z)$ and the corresponding distance modulus $\Delta m(z)$.

The conclusions are that photon physics seems to be affected by the evolution of the inhomogeneities more than by the inhomogeneities themselves. For an observer in the cheese, redshift effects are suppressed when the hole is small because of a compensating effect acting on the scale of half a hole due to spherical symmetry and to the fact that the density profile was chosen in order to have $\langle \delta\rho \rangle = 0$. This is not the case when the hole is bigger, since the evolution has more time to change the hole while the photon is going through. Moreover, as has been shown in Sec. 4.6 with Szekeres models (of which the L–T models are subclasses), small voids among overdense regions evolve more slowly than large voids do. The calculation of the angular diameter distance shows that the evolution of the inhomogeneities bends the photon paths as compared to the Friedmann case. Inhomogeneities should therefore be able, at least partly, to mimic the effects of the so-called 'dark energy'. The Friedmann model best fitting the one-big-hole Swiss-cheese universe is ΛCDM with $\Omega_m = 0.95$ and $\Omega_\Lambda = 0.05$, which is very near Einstein–de Sitter. The Friedmann model best fitting the five-small-hole Swiss-cheese model is also ΛCDM, but with $\Omega_m = 0.6$ and $\Omega_\Lambda = 0.4$, which still exhibits a non-negligible 'dark energy' component.

However, the effects of the inhomogeneities found for this model are bigger than those found by Brouzakis *et al.* (2007, 2008) and by Biswas and Notari (2008). This might be due to the fact that these authors used smaller holes with a different initial density/initial velocity profile.

In a subsequent paper, Marra *et al.* (2008) used the same Swiss-cheese toy model in order to better understand how a clumpy universe can be renormalised by a fitting procedure to give an effective ΛCDM model. The basic idea is that the observational evidences for dark energy come from the possibility that our inhomogeneous universe can be described by means of a homogeneous solution fitting the observations on our past light cone beyond some averaging scale. However, this does not imply that a primary source of dark energy exists, but only that it emerges from a phenomenological fit. If it does not exist, the observational evidence encapsulated in the Concordance model would tell us nothing else than the pure-matter inhomogeneous Universe has been merely renormalised into a ΛCDM model.

The procedure employed here to fit a phenomenological FLRW model to a Swiss-cheese model is intermediate between the fitting method of Ellis and Stoeger (1987) and an averaging approach. It is implemented on the past light cone of the observer. The physical quantities studied are the expansion and the density. The expansion behaves as in an FLRW case because of the compensation effect already found by Marra *et al.* (2007). The density behaves quite differently thanks to its sensitivity to the fact that a photon spends more and more time in the expanding large voids than in the collapsing thin high-density structures. Note, however, that this effect is independent of the one found to act on the angular diameter distance (Marra *et al.*, 2007). The best fit to an effective dark energy equation of state, quantitatively similar to the one of the Concordance model, is obtained for holes with radius 250 Mpc.

Knowing the behaviour of the density, it is possible to derive the behaviour of the Hubble parameter characterising the FLRW solution exhibiting a phenomenological source with the fitted equation of state.

Now, the insensitivity of the density to any compensation effect can yield the possibility that a Swiss cheese made of spherically symmetric holes and one where the holes slightly differ from spherical symmetry would share the same light-cone average density, thus the same redshift history for the photons and therefore the same fitted FLRW model. In this way it could be possible to go beyond the main limitation of the model studied, i.e. the assumption of spherical symmetry for the holes.

5.3.6 *Effect of inhomogeneous expansion*

The effect of inhomogeneous expansion on the luminosity distance was studied by Enqvist and Mattsson (2007). Since matter seems to be evenly distributed in the currently observed Universe, the proposed models exhibit a uniform present-day matter distribution. The observer is located at the centre of these L–T universes which are defined by two boundary condition functions:

$$H_0(r) = H + \Delta H e^{-r/r_0}, \qquad \Omega_m(r) = \Omega_0 = \text{const}, \qquad (5.67)$$

where $H_0(r) = H(t_0, r)$ is the local Hubble rate [defined as $H(t, r) \equiv R_{,t}(t, r)/R(t, r)$] at present time, $\Omega_m(r)$ is the local matter density defined by $M(r) \equiv H_0^2(r)\Omega_m(r)R_0^3(r)$ [with $R_0(r) \equiv R(t_0, r)$], and $H, \Delta H, r_0$ and Ω_0 are free parameters to be determined by the SN Ia data.

Five models are examined. In model 1, the four parameters are left free whereas, in model 2, Ω_0 is fixed to unity. The cosmological constant is included in model 3, where, to make easier the comparison with model 2, the condition $\Omega_m(r) + \Omega_\Lambda(r) = 1$ is assumed to hold [with the definition $\Omega_\Lambda \equiv \Lambda/3H_0^2(r)$]. Since the critical density depends on $H_0(r)$, the present-day matter distribution in the models with $\Omega_m(r) = \text{const}$ is not perfectly uniform. Hence, model 4 with $\rho_m(t_0, r) = \text{const}$ is also studied. Here, the boundary condition function $\Omega_m(r)$ is therefore of the form

$$\Omega_m(r) = \Omega_0(H + \Delta H)^2/(H + \Delta H e^{-r/r_0})^2. \qquad (5.68)$$

Finally, the authors consider model 5 with simultaneous Big Bang, i.e. $t_B(r) =$ const. Since this constraint leaves free only one function, Ω_m is allowed to vary with r in this model and to depend on three constant parameters, while $H_0(r)$ is left with a dependence on a fourth constant parameter H and on the varying function $\Omega_m(r)$.

For the five cases, the model parameters are computed to fit the data of the Riess *et al.* (2004) Gold Sample of 157 supernovae. In each case, a set of parameters is found to fit the supernova data with a better goodness of fit than that obtained for the standard ΛCDM model. The model giving the best fit has a perfectly uniform present-day matter density, $\Omega_m \sim 0.3$, but its Hubble parameter varies from the local value $\sim$70 km s^{-1} Mpc^{-1} to $\sim$60 km s^{-1} Mpc^{-1} over a distance scale of $\sim$500 Mpc from us.

However, as proved in Sec. 5.1, any isotropic set of observations can be fitted to appropriate inhomogeneities of an infinite number of L–T models, see also Mustapha *et al.* (1997). Anyhow, what is interesting in this work is the form of the functions which give the best fit and their physical interpretation, namely, inhomogeneities in the expansion rate coupled to a homogeneous present-day matter density. The analysis is not only in qualitative agreement with the observed homogeneity in galaxy surveys, but it gives also a value for this present-day matter density, $\Omega_m \lesssim 0.4$, compatible with its measurements.

The smallness ($\sim$15%) of the spatial variation in the Hubble parameter is another significant result of the fits in models 1 and 4. It is of the same order as the uncertainties in the model-independent determination of the local Hubble rate (Freedman *et al.*, 2001).

Various other classes of models, with different forms for the boundary condition functions $\Omega_m(r)$ and $H_0(r)$, are considered. A generic outcome is that inhomogeneities in $H_0(r)$ appear to have a much larger effect on the goodness of fit than inhomogeneities in $\Omega_m(r)$. This property can be linked to the result of Krasiński and Hellaby (2004a), reported in Sec. 4.3.2, that velocity perturbations are more efficient at generating structures than density perturbations.

The results concerning model 3 indicate that the inhomogeneities in the expansion rate and in the vacuum energy are mutually exclusive, in the sense that their combination does not lead to a better fit. A similar good fit can be achieved either by vacuum energy with no inhomogeneities, or by inhomogeneities with no vacuum energy, or by both at about half the amount of their separately deduced best-fit values. Therefore, they seem to have a very similar effect on the supernova observations.

In addition, the authors remark very accurately that, since the supernova and galaxy surveys constrain the boundary conditions only for small values of the radial coordinate r, the large r part of these conditions can be freely chosen in order to fit e.g. the CMB power spectrum.

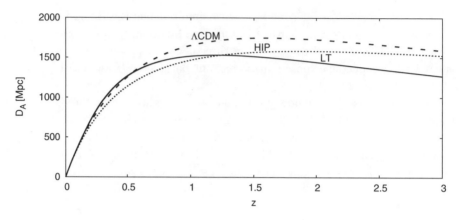

Fig. 5.4. The angular diameter distance as a function of redshift in the ΛCDM model, in a hyperbolic Friedmann model ($\Omega_m = 0.3, \Omega_\Lambda = 0$ – HIP), and in the best-fit inhomogeneous model of class I models presented in Sec. 5.3.1 with $\delta_\rho = 1.45$ and $\sigma = 0.75$ (L–T).

5.3.7 Summary of direct method investigations

The above results show that the Concordance model is not the only one which is able to describe the cosmological observations. There are other alternatives which replace dark energy with an inhomogeneity. This fact has actually been pointed out by many authors, like Dąbrowski and Hendry (1998); Célérier (2000a); Iguchi *et al.* (2002); Godłowski *et al.* (2004); Chung and Romano (2006); Alnes *et al.* (2006); Alnes and Amarzguioui (2006 and 2007); Enqvist and Mattsson (2007); Biswas *et al.* (2007); Brouzakis *et al.* (2007 and 2008); Marra *et al.* (2007); Alexander *et al.* (2007); Biswas and Notari (2008); Bolejko (2008); Enqvist (2008); Bolejko and Wyithe (2009); García-Bellido and Haugbølle (2008a, 2008b); Clifton *et al.* (2008); Yoo *et al.* (2008). [For details see a review article by Célérier (2007a) and the examples described above.] It seems that currently these two alternatives – Concordance cosmology and an inhomogeneity – are observationally indistinguishable. It is however interesting to note the observation of a 25% lack of galaxies over a $(300/h\,\mathrm{Mpc})^3$ region was reported in Frith *et al.* (2003). We anticipate that future observations will be capable of distinguishing between them.

For example, precise measurements of a maximum in the angular diameter distance would place large-scale constraints on the models (Hellaby, 2006; Araújo and Stoeger, 2009). This is depicted in Fig. 5.4 which shows the angular distance as a function of redshift in the ΛCDM model, in a hyperbolic Friedmann model, and in the best-fit void model of the first class of models presented in Sec. 5.3.1. As seen, the angular distance in an inhomogeneous model at $z > 1.5$ is distinguishable from the corresponding distance in the homogeneous models, at the level of 20%–30%. However, current estimates of the angular distance from SN Ia are predominantly at $z < 1$, and are not sufficiently precise

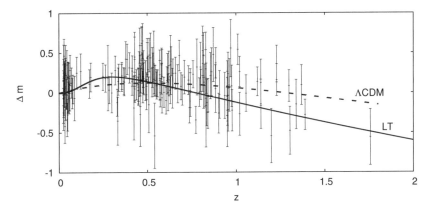

Fig. 5.5. The residual Hubble diagram for the Riess Gold Data Set, the best-fit void model of class I models presented in Sec. 5.3.1 with $\delta_\rho = 1.45$ and $\sigma = 0.75$ (L–T) and the ΛCDM model.

to discriminate between models on this basis. Current measurements of distance based on the Sunyaev–Zel'dovich effect are also too imprecise (Bonamente *et al.*, 2006). Another possible test that has been proposed in the literature is based on the time drift of cosmological redshift, i.e. a measurement of $z_{,t}$. However, this requires a very precise measurement since the typical amplitude of the effect is $\delta z \sim 10^{-10}$ on a timescale of 10 years (Uzan *et al.*, 2008). Of more immediate utility are methods based on spectral distortions of the CMB power spectrum (Goodman, 1995). These have already been used to put an upper bound of ≈ 1 Gpc for the radius of the hypothetic void (Caldwell and Stebbins, 2008). This bound is consistent with other results reported here. Of considerable promise for the future are very precise measurements of supernovae in the redshift range of 0.1–0.4 (Clifton *et al.*, 2008). In this redshift range (see Fig. 5.5) measurements of over 2000 supernovae (for example by the Joint Dark Energy Mission) will be sufficient to test the existence of such a void. However, in the near future the most promising candidate for testing inhomogeneous cosmological models are the BAO measurements. For example the WiggleZ project[6] will soon measure the dilation distance with accuracy of 2.5% at $z = 0.75$. As shown by Bolejko and Wyithe (2009), this will be sufficient to confront the hypotheses of dark energy and of the giant void.[7]

However, as stressed by Biswas and Notari (2008), there might still be two possibilities to enhance the effect of inhomogeneities, which were yet not explored. The first would be to include the possibility that structures virialise. Therefore one might expect a difference as regards the ever collapsing structure models since

---

[6] http://wigglez.swin.edu.au

[7] Note added at proof: the existence of such a large void is not at all a necessary consequence of using the L–T model to describe supernova dimming (Célérier *et al.*, 2009). Hence, all projects aimed at detecting it might be void themselves from the beginning.

these structures no longer contract after virialisation, while the void regions keep on expanding. Thus, a photon passing through both condensed structures and voids might experience a comparatively larger redshift. The second possibility would be to give up on the proper matching, to allow for the movement of the boundaries of the inhomogeneous patches. This would leave open the possibility that significant backreaction effects could appear in a more general case.

Moreover, we have seen that, within L–T Swiss cheese, inhomogeneities in the expansion and structure evolution have a larger effect on the apparent dimming of SN Ia than density contrast. Structure evolution analyses realised in the framework of the quasi-spherical Szekeres models with no symmetry and reported in Sec. 4.6 could be therefore of interest. It was found that, when comparing the evolution of a double structure in quasi-spherical Szekeres models with that of single structures in L–T models, the growth of the density contrast was 5 times faster in the former than in the latter. We can therefore suspect an enhanced effect on the SN Ia dimming by nonspherical structures. It was also found that large isolated voids evolve much more rapidly than voids surrounded by high overdensities. The former could be thus more efficient at enhancing the supernova dimming.

5.4 The inverse problem

The inverse problem is much more involved than the direct one since we know that to every $D_L(z)$ function can correspond an infinite number of L–T solutions. This is the reason why the authors who have tried to deal with this issue have, most of the time, added some a priori constraint to the model to make the problem tractable. Moreover, all the attempts found in the literature consider only the central observer case.

A first attempt was made by Iguchi *et al.* (2002) for two restrictive cases: that of a pure bang time inhomogeneity ($E(r) = $ const) and that of a pure curvature inhomogeneity ($t_B(r) = 0$).

5.4.1 The flat case

In this section, we summarise the work done by Vanderveld *et al.* (2006) devoted to determining the bang time parameter functions $t_B(r)$ defining a subclass of the $E(r) = 0$ class of L–T models with a central observer yielding the luminosity distance measured from the SN Ia observations. This is reported in their Sec. III where they deal with a reduced setting of the inverse problem, for this particular case. (We note that this method is different from the one they used in their Sec. II.C, where they study a subsample of the same $E(r) = 0$ class of L–T models with an a priori guessed form for the bang time function, to see if a quantity defined as an 'apparent' deceleration parameter can be negative in some region around the centre where the observer is located. However, we have seen in Sec. 5.1.4 that this was not the point and that one must not focus on the 'acceleration' supposedly seen in the data.) We give below the main steps of the reasoning, referring the interested reader to the original article.

The authors first recall the expression for the angular diameter distance in $E = 0$ L–T models, which is

$$D_A(r) = R(t_n(r), r) = rT^2(t_n(r), r), \tag{5.69}$$

where $T \equiv [t - t_B(r)]^{1/3}$ and where they have specialised to units in which $6\pi G\rho_0(t_0) = 1$ (here, the subscript 0 refers to the position of the central observer in space and time). They also define the equivalent Friedmann radial coordinate to be

$$r_F(z) \equiv (1 + z)D_A(z), \tag{5.70}$$

in terms of which they obtain

$$rT^2(1 + z) = r_F(z) = \frac{D_L(z)}{1 + z}. \tag{5.71}$$

The problem can be set therefore as follows: from the SN Ia observations, interpreted in a Friedmannian framework, one is given equivalently $r_F(z)$, $D_L(z)$ or $D_A(z)$, and one wants to find the corresponding zero energy, zero cosmological constant L–T model, i.e. the corresponding bang time function t_B.

The equations defining such models may be written

$$\frac{dT}{dr} = -\frac{1}{3} + \frac{dt_B}{dr}\left(\frac{2r}{9T^3} - \frac{1}{3T^2}\right) \tag{5.72}$$

and

$$\frac{1}{(1 + z)}\frac{dz}{dr} = \left(\frac{2}{3T} + \frac{2r}{9T^4}\frac{dt_B}{dr}\right). \tag{5.73}$$

Combining (5.72) and (5.73), one obtains

$$\frac{dt_B}{dz} = \frac{3T}{2(1 + z)(r/T - 1)}\frac{d}{dz}[T^2(1 + z)] \tag{5.74}$$

and

$$\frac{r}{T}\frac{dT}{dz} + \left(\frac{r}{T} - 1\right)\frac{dr}{dz} = \frac{1}{1 + z}\left(r - \frac{3T}{2}\right). \tag{5.75}$$

Defining $X \equiv T^2(1 + z)$, (5.74) and (5.75) are recast into

$$\frac{1}{X}\frac{dX}{dz} = \frac{r_F\sqrt{1 + z}/X^{3/2} - 1}{r_F(3/2 - r_F\sqrt{1 + z}/X^{3/2})}\left[\frac{3X^{3/2}}{2(1 + z)^{3/2}} - \frac{dr_F}{dz}\right] \tag{5.76}$$

and

$$\frac{dt_B}{dz} = \frac{3X^{3/2}}{2r_F(1 + z)^{3/2}(3/2 - r_F\sqrt{1 + z}/X^{3/2})}$$
$$\times \left[\frac{3X^{3/2}}{2(1 + z)^{3/2}} - \frac{dr_F}{dz}\right]. \tag{5.77}$$

Given the expression for $r_F(z)$, (5.76) is a first-order ordinary differential equation for $X(z)$. Solving this equation for X, then plugging the result into (5.77) yields a first-order ordinary differential equation for $t_B(z)$. Both equations become singular when

$$\frac{r_F(z)\sqrt{1+z}}{X^{3/2}} = \frac{3}{2}. \tag{5.78}$$

Solutions $z = z_{crit}$ of (5.78), if they exist, are critical points of the differential equations (5.76) and (5.77). Near the critical point,

$$\frac{1}{X}\frac{dX}{dz} \approx \frac{1}{2(3/2 - r_F\sqrt{1+z}/X^{3/2})}\left[\frac{1}{1+z} - \frac{d\ln r_F}{dz}\right] \tag{5.79}$$

and

$$\frac{dt_B}{dz} \approx \frac{r_F(z)}{(1+z)(3/2 - r_F\sqrt{1+z}/X^{3/2})}\left[\frac{1}{1+z} - \frac{d\ln r_F}{dz}\right]. \tag{5.80}$$

Equations (5.79) and (5.80) show that transcritical solutions, nonsingular at the critical point, are possible provided $(1+z)d\ln r_F/dz = 1$ at the critical point. However, for a general function $r_F(z)$, the conditions for passing smoothly through the critical point might not be generically satisfied, and both $d\ln X/dz$ and dt_B/dz might diverge there.

The authors show therefore that a flat L–T model can mimic a generic observed r_F only up to some limiting redshift z_0 below z_{crit}. However, it is possible to find nonpathological transcritical solutions. Trying to construct such models within a special class of flat L–T solutions, a priori chosen for their nice theoretical properties, the authors fail to reproduce the observational constraints.

In fact, these authors are not the first to conclude the inverse problem has a limiting redshift that you can't get past. But there must be a solution, since even an FLRW model has such a singularity in its inverse problem, and yet a complete exact solution is known. This problem was solved in Lu and Hellaby (2007) by using a series solution of the differential equations near the maximum in the areal radius, and Hellaby (2006) and McClure and Hellaby (2008) used it to advantage, showing it provides extra observational information. (See a discussion of this work at the end of this section.)

Moreover, it must be stressed that a limiting redshift above which the solution is no longer valid is not a drawback from the coincidence point of view. It might on the contrary be considered as a nice property of such a model, since what is desired is also a transition from accelerated-like to non-accelerated-like expansion at some redshift which might also be the location of a transition from small-scale inhomogeneity to large-scale homogeneity where the L–T solution could match smoothly to a Friedmann one with $t_B(r) \rightarrow$ const.

We stress that Vanderveld $et\ al.$'s conclusions apply only to flat L–T models and that the case of nonflat ones will be studied in the sections below. Also, since their analysis does not consider the fully general case, they do not exclude the possibility that a transcritical model agreeing with the observations may exist.

5.4.2 A map from $\{E(r), D_L(z), R_0(r)\}$ to $M(r)$

Chung and Romano (2006) derived a map from $\{E(r), D_L(z), R_0(r)\}$ to $M(r)$, where $R_0(r) \equiv R(t_0, r)$ is an initial condition specification for $R(t, r)$ and t_0 is not necessarily today. It is a generalisation to $E(r) \neq 0$ of the similar inverse problem considered above. But here the function sought is not $t_B(r)$ but $M(r)$. This inversion method has the advantage that the physical observable $D_L(z)$ can be mapped to the geometry of the underlying model directly, without having to guess $M(r)$. Therefore, if one knows the radial geodesic history of a single photon which was emitted at event (t_1, r_1) and observed at (t_2, r_2), one knows the full spacetime geometry of the L–T model in the region $(t_1 < t < t_2, r_1 < r < r_2)$, owing to its spherical symmetry. The method is as follows.

Starting from (5.17), (3.3) and (3.4), giving the luminosity distance in L–T models as derived in Célérier (2000a), one can simplify (3.3) and (3.4) in the regime where $R_{,t}$, the 'local expansion' rate, maintains the same sign by rewriting (2.2) (with no cosmological constant) as

$$R_{,t} = \ell\sqrt{2E(r) + \frac{2M(r)}{R(t,r)}}, \tag{5.81}$$

where $\ell \equiv \pm 1$ specifies whether there is local expansion or contraction. Since the solution to this differential equation requires a specification of one of the functions of r on the initial time hypersurface, the authors choose that function to be $R_0(r) \equiv R(t_0, r)$. Differentiating this equation with respect to r, one obtains an expression for $R_{,tr}$ which reads

$$R_{,tr} = \ell\frac{E_{,r} + M_{,r}/R - MR_{,r}/R^2}{\sqrt{2E + 2M/R}}. \tag{5.82}$$

Inserting this expression into (3.3) and (3.4), one obtains a set of two differential equations from which every second-order derivative has disappeared and which is written

$$\frac{dr}{dz} = \frac{\ell\sqrt{(1 + 2E)(2E + 2M/R)}}{(1 + z)(E_{,r} + M_{,r}/R - MR_{,r}/R^2)}, \tag{5.83}$$

$$\frac{dt}{dz} = \frac{-\ell|R_{,r}|\sqrt{2E + 2M/R}}{(1 + z)(E_{,r} + M_{,r}/R - MR_{,r}/R^2)}. \tag{5.84}$$

In Appendix B of their article, the authors give two exact solutions of (2.2) (without cosmological constant), for each case $E > 0$ and $E < 0$, respectively their Eqs. (B1) and (B2). To find an expression for $R_{,r}$ as a function of z, they differentiate these equations with respect to r and obtain a linear equation for

$R_{,r}$, which can be solved to find $R_{,r}$ as a function of $R, R_0, E, M_1, R_{,r0}, E_{,r}, M_{,r}$ and t, with $M_1(z) \equiv M[r(z)]$ [their Eq. (14) in the $E > 0$ case]. We do not write here the full equations since this would be rather long and we refer the interested reader to the original article.

Then, one replaces the function $M_{,r}(r)$ in (5.83) and (5.84) by

$$M_{,r}(r) = \frac{dM_1}{dz}\left(\frac{dr}{dz}\right)^{-1}. \tag{5.85}$$

Since there are three unknown functions, $M_1(z), r(z)$ and $t(z)$, and (5.83) and (5.84), with appropriate substitutions for $R_{,r}$ and M, provide only two independent equations, one needs a third equation which is given by dR/dz through the chain rule:

$$\frac{d}{dz}R = \ell\sqrt{2E + \frac{2M_1}{R}}\frac{dt}{dz} + R_{,r}\frac{dr}{dz}. \tag{5.86}$$

Therefore, for a given set of $\{E(r), D_L(z), R_0(r)\}$, the set of differential equations (5.83), (5.84) and (5.86) can be solved for $\{t(z), r(z), M_1(z)\}$. Finally, to obtain $M(r)$, one inverts $r(z)$ which gives

$$M(r) = M_1[z(r)]. \tag{5.87}$$

In practice, the procedure is a bit more difficult to implement because the differential equations can become singular for certain choices of $\{E(r), D_L(z), R_0(r)\}$ and also for the initial conditions (see the original article for a discussion). However, the authors solve this set for several examples to demonstrate the accuracy of the method.

In using this map to implement the chosen examples, the authors find that the luminosity distance is typically, depending on the choice of $E(r)$, an effective probe of the L–T geometry only for $z \lesssim 1$, since $D_L(z)$ has a universal behaviour in the limit $(R_0 = 0, E \to 0)$. More precisely, $D_L(z)$ is fixed independently of $M(r)$ for $(R_0 = 0, E = 0)$, since that limit corresponds to the $\Omega_m = 1$ Friedmann universe. A negative consequence of this feature is that the differential equation map from $\{E(r), D_L(z), R_0(r)\}$ to $M(r)$ fails to be numerically accurate beyond $z \sim 1$. Nonetheless, the authors show how the numerical solution can be patched to a semi-analytic one beyond $z = 1$ to obtain a good fit (to within around 5% for the redshifts of interest) with null cosmological constant L–T cosmologies reproducing the luminosity distance function of a Friedmann model with $\Omega_\Lambda = 0.7$ and $\Omega_m = 0.3$.

To reduce the degeneracy of the mapping, Romano (2007b) proposed to introduce another function of the redshift, called the 'radial spherical shell energy', $\rho_{SS}(z)$. However, this quantity can be used for this purpose only provided one can relate it to some astrophysical observable. A natural candidate could be the galaxy number count, $n(z)$, but its relation to $\rho_{SS}(z)$ is rather complicated and implies bringing in another function of the redshift, $F(z)$, to take into account different observational corrections and bias: $n(z) = F(z)\rho_{SS}(z)$. Thus, $F(z)$ would add, in principle, a further degeneracy to the problem, as properly stressed by the author himself.

5.4.3 Expansion fitting

Another method proposed by Tanimoto and Nambu (2007) to deal with the inverse problem in the spirit of Célérier (2000a) is to expand the luminosity distance expression for the L–T solution, $D_L(z)$, in powers of the redshift z and compare the series coefficients to those of a similar expansion of the observed luminosity distance, $D_L^{(in)}(z)$. Even if, in the end, the degeneracy of the problem forces one to make an arbitrary choice of one of the coefficients defining the parameter functions, such a constraint is not set here a priori. Therefore, this method can be applied to models with various physical constraints.

The authors chose to work in the 'light cone gauge', where the radial coordinate r is defined so that the light rays caught by the central observer at $(t, r) = (t_0, 0)$ are expressed by

$$t = t_0 - r. \tag{5.88}$$

In this case $dr/dt = -1$ must hold along those light rays, and, from the metric of (2.1), it is equivalent to imposing

$$\frac{R_{,r}(t_0 - r, r)}{\sqrt{1 + 2E(r)}} = 1. \tag{5.89}$$

This still leaves two degrees of freedom [amongst $E(r)$, $M(r)$ and $t_B(r)$] with which to fit observational data. Together with the general expression for R given by (2.126), this gauge choice makes the expansion systematic and straightforward.

The procedure runs therefore as follows. First, R is expanded in powers of r from (2.126) where $R = R(t_0 - r, r)$ has been inserted. The result is converted into an expansion in powers of z with the use of (3.3) where t is replaced by its expression given by (5.88). To eliminate the so-called 'weak singularity' at the origin,

$$\rho_{,r}(t, 0) = 0 \tag{5.90}$$

is imposed (see Sec. 2.1.2). Now, the three free functions are expanded in powers of r as follows:

$$E(r) = \frac{e_2}{2}r^2 + \frac{e_3}{3!}r^3 + \frac{e_4}{4!}r^4 + \mathcal{O}(r^5),$$

$$t_B(r) = b_1 r + \frac{b_2}{2}r^2 + \mathcal{O}(r^3),$$

$$M(r) = \frac{m_3}{3!}r^3 + \frac{m_4}{4!}r^4 + \frac{m_5}{5!}r^5 + \mathcal{O}(r^6), \tag{5.91}$$

where the coefficients $e_i \equiv E^{(i)}(0), b_i \equiv t_B^{(i)}(0)$ and $m_i \equiv M^{(i)}(0)$ are constants, with $m_3 > 0$ $[f^{(i)}(0)$ meaning $d^i f/dr^i$ taken at $r = 0]$. The next step is to express the expansion of $D_L(z)$ in terms of these coefficients.

Prior to the general computation, a partial calculation at first order for the expansion of $D_L(z)$ allows one to determine the Hubble constant from the known result $D_L(z) = z/H_0 + \mathcal{O}(z^2)$ (Célérier 2000a), which gives

$$H_0 = \frac{2}{3t_0}\left[1 + x_0 \frac{S_{,x}(x_0)}{S(x_0)}\right], \quad x_0 \equiv -e_2\left(\frac{t_0}{m_3}\right)^{2/3}, \tag{5.92}$$

where S is the function defined and discussed in Sec. 2.1.8.

Then, the calculations are developed to the higher orders. Since these calculations are rather long and straightforward, we do not write them down here and we refer the interested reader to the original article. The final form of $D_L(z)$ is

$$
\begin{aligned}
D_L(z) ={}& \frac{z}{H_0} + \frac{1}{4H_0}\left(1 + \frac{e_2}{H_0^2}\right)z^2 \\
&- \frac{1}{48H_0}\left[\frac{1 - e_2/H_0^2}{1 - H_0 t_0}\left(13 + \frac{2e_2}{H_0^2} + \frac{12b_2}{H_0}\right) - 33 + \frac{e_2}{H_0^2}\left(7 - \frac{4e_2}{H_0^2}\right)\right. \\
&+ \left.\frac{1}{1 - H_0 t_0}\left(H_0 t_0 + \frac{2 - 3H_0 t_0}{e_2/H_0^2}\right)\frac{e_4}{H_0^4}\right]z^3 + \mathcal{O}(z^4),
\end{aligned}
\tag{5.93}
$$

with constraints upon the coefficients, obtained in the course of the calculations, which read

$$S(x_0) = t_0^{-2/3} m_3^{-1/3}, \tag{5.94}$$

$$\frac{e_2}{H_0^2} = 1 - \frac{m_3}{3H_0^2}, \tag{5.95}$$

$$m_4 = 6H_0 m_3, \tag{5.96}$$

$$b_1 = 0, \quad e_3 = \frac{e_2 m_4}{3m_3} = 3H_0 e_2, \tag{5.97}$$

$$m_5 = \frac{m_3}{2}\left[15H_0^2 + 5\frac{e_4}{e_2} + \frac{5}{3(1 - H_0 t_0)}\left(13H_0^2 + 12H_0 b_2 + 2e_2 - \frac{e_4}{e_2}\right)\right]. \tag{5.98}$$

It is now easy to determine an L–T model whose $D_L(z)$ coincides with a given luminosity distance function $D_L^{(in)}(z)$, at least up to the third order. The input function is expanded as

$$D_L^{(in)}(z) = I_1 z + \frac{I_2}{2}z^2 + \frac{I_3}{3!}z^3 + \mathcal{O}(z^4). \tag{5.99}$$

A comparison with (5.93) gives immediately

$$H_0 = \frac{1}{I_1},$$

(5.100)

$$\frac{e_2}{H_0^2} = 2\frac{I_2}{I_1} - 1,$$

(5.101)

$$\frac{1}{1 - H_0 t_0}\left(H_0 t_0 + \frac{2 - 3H_0 t_0}{e_2/H_0^2}\right)\frac{e_4}{H_0^4} + \frac{1 - e_2/H_0^2}{1 - H_0 t_0}\frac{12 b_2}{H_0}$$
$$= -8\frac{I_3}{I_1} + 33 - \frac{e_2}{H_0^2}\left(7 - \frac{4e_2}{H_0^2}\right) - \frac{1 - e_2/H_0^2}{1 - H_0 t_0}\left(13 + \frac{2e_2}{H_0^2}\right).$$

(5.102)

Here H_0 is given by (5.100), and this constrains the parameters e_2, m_3 and t_0 through (5.92). The coefficient e_2 follows from (5.101), which in turn yields m_3 through (5.95). Then, the current time t_0 for the observer is implicitly determined from (5.94) with the help of the values of e_2 and m_3. The parameters e_2, m_3 and t_0 can now be calculated for given values of I_1 and I_2. However, adding a given value for I_3, we can see that (5.102) represents only one constraint for two variables e_4 and b_2. This comes from the fact, already stressed in Sec. 5.1, that the problem of matching a $D_L^{(in)}(z)$ function to a given L–T model is highly degenerate in that multiple inequivalent models can give rise to the same $D_L(z)$. However, if one of the parameters e_4 or b_2 can be chosen according to other considerations, then the other parameters follow from (5.96) to (5.98).

Note that, due to the condition $m_3 > 0$, this process is solvable only for $I_2/I_1 < 1$.

5.4.4 Extracting the cosmic metric from observations

The work discussed above focusses on reproducing the SN Ia luminosity observations. Since the L–T model has 2 free physical functions, and the variation of the luminosity distance with z only fixes one of them, it is necessary to specify the second function arbitrarily, e.g. $t_{B,r} = 0$.

Papers by Lu and Hellaby (2007) and McClure and Hellaby (2008) initiate a programme to extract metric information from observations. This is the full inverse problem, and is not degenerate. This work is closely allied to the observational cosmology programme initiated by Kristian and Sachs (1966) and pursued by e.g. Ellis *et al.* (1985); Stoeger *et al.* (1992); Maartens *et al.* (1996); Araújo *et al.* (2001); Araújo *et al.* (2008); Hellaby and Alfedeel (2009) amongst others. To date it has assumed the metric has the L–T form, as a relatively simple case to start from, though the long-term intention is to remove the assumption of spherical symmetry. These papers develop and code an algorithm that generates the L–T metric functions, given observational data on the redshifts, apparent luminosities or angular diameters, and number counts of galaxies, as well as estimates for the absolute luminosities or true diameters and source masses, as functions of z. These data give us $R_n(z)$ and $\mu n(z)$, where R_n is the diameter distance of (5.16), n is the density of sources in redshift space, and μ is the mean mass per source. They allow both of the physical functions of an L–T model to be determined. In this context, a method for testing theories of source evolution was proposed by Hellaby (2001).

The basic algorithm is that outlined by Mustapha, Hellaby and Ellis (1997). The gauge choice $\mathrm{d}t_n/\mathrm{d}r = -1$ fixes the r coordinate, giving the past null cone as $t_n = t_0 - r$, and then the coordinate r and the L–T functions M and $W = \sqrt{1 + 2E}$ are determined from the equations

$$\frac{\mathrm{d}r}{\mathrm{d}z} = \phi, \tag{5.103}$$

$$\frac{\mathrm{d}\phi}{\mathrm{d}z} = \phi \left\{ \frac{1}{1+z} + \frac{\mathrm{d}^2 R_n/\mathrm{d}z^2 + \kappa\mu n\phi/(2R_n)}{\mathrm{d}R_n/\mathrm{d}z} \right\}, \tag{5.104}$$

$$\frac{\mathrm{d}M}{\mathrm{d}z} = \frac{\kappa\mu n W}{2}, \tag{5.105}$$

$$W = \frac{\mathrm{d}R_n/\mathrm{d}z}{2\phi} + \frac{\left(1 - 2M/R_n - \Lambda R^2/3\right)\phi}{2\mathrm{d}R_n/\mathrm{d}z}, \tag{5.106}$$

where n is the density of sources in redshift space, and μ is the mean mass per source. The bang time t_B follows from (2.4) via

$$t_B = t_n - \int_0^{R_n} \frac{\mathrm{d}R}{R_{,t}}, \tag{5.107}$$

or one may use (2.5a)–(2.7b).

However, two regions require a different treatment, in order to avoid numerical singularities in the basic differential equations; the origin, and the maximum in the areal radius on the past null cone $R_n = R_m$. At the maximum, where we write $M = M_m$, the relation

$$3R_m - 6M_m - \Lambda R_m^3 = 0 \tag{5.108}$$

must obtain, giving just $R_m = 2M_m$ when $\Lambda = 0$. In addition, because the observational data consist of many discrete sources, whereas the metric functions are continuous, the data have to be binned and averaged. Lu and Hellaby (2007) constructed the initial numerical procedure, and successfully tested it using fake data generated from a variety of homogeneous and inhomogeneous cosmological models. McClure and Hellaby (2008) investigated the effect of noisy observational data, improving several aspects of the numerical procedure so that it can handle input data that are subject to statistical fluctuations. They verified that, as pointed out by Hellaby (2006), the data at R_m provide a consistency check on the numerical results at that locus, which can be used to detect systematic errors, thus allowing a partial correction that makes the result self-consistent.

They extended the numerical code to provide estimates of the uncertainties in the metric functions, and they demonstrated that the procedure is stable for low z, but beyond R_m, around $z > 1.5$, there seems to be a stability problem in the mass/energy function that grows with z. It is not yet clear if this is specific to the method used, or is a generic feature of the cosmological inverse problem.

6
The horizon problem

6.1 Horizon problem and inflation

In hot Big Bang models, the comoving region over which the CMB is observed
to be homogeneous to better than one part in 10^5 at the last-scattering surface is
much larger than the intersection of this surface with the future light cone from
any point p_B of the Big Bang. Since this light cone provides the maximal distance
over which causal processes could have propagated from p_B, the observed quasi-
isotropy of the CMB remains unexplained. As shown by Célérier (2000b), this
'horizon problem' develops sooner or later in any cosmological model exhibiting
a spacelike singularity such as that occurring in the standard FLRW universes.

Even inflation, which was put forward in order to remove this drawback in the
framework of standard homogeneous cosmology, only postpones the occurrence
of the horizon problem, since it does not change the spacelike character of the
singularity and is insufficient to solve it permanently. This is shown in Fig. 6.1,
where thin lines represent light cones and where the CMB as seen by an observer
O corresponds to the intersection of the observer's backward light cone with the
last-scattering line. For a complete causal connection to occur between every
pair of points in this intersection segment, all backward light signals issuing from
points therein must reach the vertical axis before they reach the spacelike Big
Bang curve. The event L is thus a limit beyond which any observer experiences
the horizon problem. Adding an inflationary phase in the primordial history
of the universe amounts to adding a slice of de Sitter spacetime, indicated here
by the region between the dashed line and the Big Bang. The effect of this
region is merely to postpone the event L, allowing the current observer O to see
a causally connected CMB. At later times, when the observer reaches the region
above L, either the Universe is found to be inhomogeneous (see Sec. 6.2), or the
horizon problem appears again.

6.2 Permanent solution

A proposal at variance with inflation was put forward, solving the horizon problem by means of a shell-crossing singularity in a spherically symmetric inhomogeneous model (Célérier and Schneider, 1998; Célérier, 2000b; Célérier and
Szekeres, 2002). This allows a permanent solution to the problem for all observers
regardless of their spacetime location.

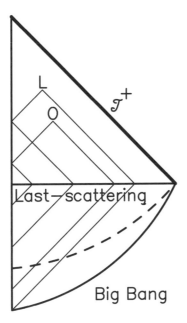

Fig. 6.1. Carter–Penrose diagram showing the horizon problem in a universe with spacelike singularity (owing to spherical symmetry, we show only half of the diagram).

A property of a large class of shell crossings, valid for general spherically symmetric models, is that they are timelike. A derivation, generalising a line of reasoning first proposed by Hellaby and Lake (1985), was developed by Célérier and Szekeres (2002) and is improved here. It runs as follows.

The general spherically symmetric line element can be written

$$ds^2 = B^2(r,t)dt^2 - A^2(r,t)dr^2 - R^2(r,t)(d\vartheta^2 + \sin^2\vartheta d\varphi^2). \qquad (6.1)$$

A typical shell-crossing surface appears at $t = b(r)$, where

$$A = [t - b(r)]^a f(r,t) = 0, \qquad B \neq 0, \qquad R \neq 0. \qquad (6.2)$$

The normal n_α to the surface $A =$ const (here $A = 0$), is

$$n_\alpha \propto (A_{,t}, A_{,r}, 0, 0). \qquad (6.3)$$

With the metric of (6.1), the squared norm of this normal vector is

$$n_\alpha n^\alpha = \frac{A_{,t}^2}{B^2} - \frac{A_{,r}^2}{A^2}. \qquad (6.4)$$

According to (6.2), if $a > 1/2$, then $A^2 = 0$, and, if $A_{,r} \neq 0$, the above expression for $n_\alpha n^\alpha$ is negative implying a timelike shell-crossing surface.

Now, the Carter–Penrose diagram of Fig. 6.2 shows that a non-spacelike singularity always gives rise to an everywhere causally connected model of the

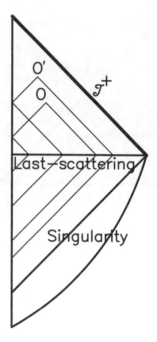

Fig. 6.2. Carter–Penrose diagram showing causal connectedness of a universe with non-spacelike singularity.

Universe. Every pair of points in the CMB seen by the current observer O is causally connected since a past light signal from any point in the segment of the last-scattering surface seen by O reaches the vertical axis before arriving at the null (straight line) or timelike (curved line) singularity. The same holds for any event O' in the observer's past or future.

6.3 The Delayed Big Bang (DBB) model

To solve the horizon problem in the manner indicated in Sec. 6.2, we need a cosmological model which exhibits a non-spacelike singularity that is encountered when travelling backward on the observer's light cone. Since we have shown that shell crossings in a large class of spherically symmetric models are timelike, we choose to work within such a class of models. For simplicity, we employ L–T models, because they exhibit the needed symmetry and allow a fully analytical exact reasoning (Célérier and Schneider, 1998).

To deal with the horizon problem, one has to consider light cones. As was shown by Célérier and Schneider (1998), a large class of L–T models can be found for which the horizon problem is solved by means of light cones never leaving the matter-dominated era. This validates the use of these models. Moreover, since there is evidence that the observed Universe does not present appreciable spatial curvature on very large cosmological scales, it will be here approximated by the flat $E(r) = 0$ L–T model.

The function $M(r)$ is used to define the radial coordinate r: $M(r) \equiv M_0 r^3$, where M_0 is a constant. With the above definition for r, the analytical expression for $R(t, r)$ given by (2.6) becomes

$$R(t, r) = \left(\frac{9GM_0}{2}\right)^{1/3} r[t - t_B(r)]^{2/3}. \tag{6.5}$$

In this case, the g_{rr} component of the metric tensor is

$$R_{,r}(t, r) = \left(\frac{9GM_0}{2}\right)^{1/3} [t - t_B(r)]^{-1/3}[t - t_B(r) - \frac{2}{3}rt_{B,r}(r)]. \tag{6.6}$$

From (6.2) and the reasoning following that equation, we see that the shell-crossing surface

$$t - t_B(r) - \frac{2}{3}rt_{B,r}(r) = 0 \tag{6.7}$$

is timelike since here $a = 1 > 1/2$.

On the other hand, one can always choose $t_B(0) = 0$ at the centre $(r = 0)$ of symmetry by an appropriate translation of t.

Equation (6.5) substituted into (2.3) gives

$$\rho(t, r) = \frac{1}{2}\pi G[3t - 3t_B(r) - 2rt_{B,r}(r)]^{-1}[t - t_B(r)]^{-1}. \tag{6.8}$$

This expression for ρ leads to two potentially undesirable consequences:

(1) The energy density goes to infinity on the Big Bang surface, $t = t_B(r)$, and on the shell-crossing surface (6.7).

(2) This energy density presents negative values in the region of the Universe located between the shell crossing (6.7) and the Big Bang singularity, and corresponding to $3t - 3t_B(r) - 2rt_{B,r}(r) < 0$ and $t - t_B(r) > 0$. Of course, a negative energy density is unphysical for dust.

Since a shell crossing is usually considered to be a deficiency of L–T models one generally tries to avoid it by setting, for our $E(r) = 0$ case: $t_{B,r}(r) \leq 0$ and $M_{,r}(r) \geq 0$ in the region where $R_{,r} > 0$ (Hellaby and Lake, 1985).

But, as will be developed further on, we need an increasing Big Bang function $t_B(r)$ to solve the horizon problem. It is possible to circumvent this difficulty.

(1) A way out the shell-crossing surface problem might be to consider that, as the energy density increases while reaching its neighbourhood from higher values of t, radiation becomes the dominant component of the Universe, pressure can no more be neglected and the L–T model no longer holds.

(2) However, since the light cones of interest, which are null surfaces, never leave the region situated above the timelike shell-crossing surface, where $3t - 3t_B(r) - 2rt_{B,r}(r)$ remains always positive, and $t - t_B(r) > 0$ by construction, therefore the energy density of (6.8) remains positive.

One of the key points of the reasoning here proposed is a shell-crossing surface situated above the Big Bang surface, its time coordinate monotonically increasing with r to let the light cones remain in the matter-dominated domain. This is always the case if

$$t_B(r = 0) = 0,$$

$$t_{B,r}(r) > 0 \qquad \text{for all } r \neq 0,$$

$$t_{B,r}(r = 0) = 0, \tag{6.9}$$

$$5t_{B,r}(r) + 2rt_{B,rr}(r) > 0 \qquad \text{for all } r,$$

$$rt_{B,r}|_{r=0} = 0.$$

The fourth condition of (6.9) implies $A_{,r} \neq 0$ and therefore a timelike shell-crossing surface as shown in Sec. 6.2. Note that we have added to the specifications of Célérier and Schneider (1998) and Célérier (2000b) the constraint that $t_{B,r}$ should vanish at $r = 0$ to allow the model to satisfy the no-central-singularity conditions in order to get a well-behaved model whatever the observer's position. With these specifications, the physical singularity of the model – i.e. the first surface encountered on a backward path from the observer where the energy density and the scalar curvature go to infinity – is the shell-crossing surface. Therefore, the region between the Big Bang $t = t_B(r)$ and the shell-crossing surface, containing the unwanted negative energy density, is excluded from the part of the model intended to describe the matter-dominated region of the Universe.

We have thus constructed a class of models permanently solving the horizon problem with no need for any inflation phase. A $t_B(r)$ function increasing with r implies that the Big Bang 'occurred' at later cosmic time t for larger r, hence the evocative 'Delayed Big Bang' chosen to qualify this singularity.

7

CMB temperature fluctuations

7.1 Light propagation effects

The last-scattering surface is the most remote region which is observable using electromagnetic radiation. Since photons on their way pass through large-scale inhomogeneities such as voids, clusters and superclusters, it is important to know how the light propagation phenomena affect the CMB radiation. In the standard approach the CMB temperature fluctuations are analysed by solving the Boltzmann equation within linear perturbations around the homogeneous and isotropic FLRW model (Seljak and Zaldarriaga, 1996; Seljak *et al.*, 2003).[1] The use of the FLRW metric for the background model results in a remarkably good fit to the CMB data (Hinshaw *et al.*, 2009). However, the assumption of homogeneity, which is also consistent with other types of cosmological observations, is not a direct consequence of them (Ellis, 2008). It is often said that such theorems as the Ehlers–Geren–Sachs (1968) theorem and the 'almost EGS theorem' (Stoeger *et al.*, 1995), justify the application of the FLRW models. These theorems state that if anisotropies in the cosmic microwave background radiation are small for all fundamental observers, then locally the Universe is almost spatially homogeneous and isotropic. The founding assumption of these theorems, namely the local Copernican principle applied to the 'U region', i.e. 'the region within and near our past light cone from decoupling to the present day', is not mandatory and we have already stressed it needs still to be tested. Moreover, as shown by Nilsson *et al.* (1999), it is possible that the CMB temperature fluctuations are small but the Weyl curvature is large.

In such a case the geometry of the Universe is far from the Robertson–Walker geometry, and the applicability of FLRW models is not justified. Moreover, the applicability of the linear approach can be questionable, since the density contrast within cosmic structures is much larger than unity. Therefore, there is a need for the application of exact inhomogeneous models to a study of light propagation and its impact on CMB temperature fluctuations. This issue has been extensively studied within spherically symmetric models – within the thin shell approximation (Thompson and Vishniac, 1987; Inoue and Silk, 2006, 2007) and within the Lemaître–Tolman model (Panek, 1992; Arnau *et al.*, 1993; Saez

---

[1] This approach is implemented in such codes like CMBFAST (http://www.cfa.harvard.edu/~mzaldarr/CMBFAST/cmbfast.html), CAMB (http://www.camb.info/) or CMBEASY (http://www.cmbeasy.org/).

et al., 1993; Arnau et al., 1994; Fullana et al., 1994; Rakić et al., 2006; Masina and Notari, 2009). However, cosmic structures are not spherically symmetric. Therefore, in this section we will focus on the study of light propagation in the Swiss-cheese Szekeres model.

7.1.1 Temperature fluctuations

Assuming that the black body spectrum is conserved during the evolution of the Universe, the temperature must be proportional to $1 + z$:

$$\frac{T_e}{T_o} = 1 + z. \tag{7.1}$$

Then from (7.1) the temperature fluctuations measured by a comoving observer are:

$$\left(\frac{\Delta T}{T}\right)_o = \frac{T_e/(1+z) - \overline{T}_e/(1+\overline{z})}{\overline{T}_e/(1+\overline{z})}, \tag{7.2}$$

where quantities with overbars refer to the average quantities, i.e. the quantities obtained in the homogeneous Friedmann model.

Let us write the temperature at emission as $T_e = \overline{T}_e + \Delta T_e$. Then (7.2) becomes:

$$\left(\frac{\Delta T}{T}\right)_o = \frac{\overline{z} - z}{1 + z} + \left(\frac{\Delta T}{\overline{T}}\right)_e \frac{1 + \overline{z}}{1 + z}. \tag{7.3}$$

As can be seen from (7.3), the observed temperature fluctuations on the CMB sky are caused by the light propagation effect (first term) and by the temperature fluctuation at the decoupling instant (the second term).

To calculate the light propagation accurately, without assumptions such as small density contrast and linear evolution, a model of cosmic structures based on exact solutions of the Einstein field equations is needed. Unfortunately, there are only a few exact inhomogeneous cosmological models, and in none of them can the cosmic web-like structure be described. This obstacle can be overcome by using many inhomogeneous models of cosmic structures and joining them in a Friedmann background to obtain an inhomogeneous patchwork model of the Universe.

7.1.2 Model specification and evaluation

Our Swiss-cheese model consists of a variety of inhomogeneous Szekeres patches or regions in a Friedmann background. In fact, they are arranged so that the light rays followed always pass directly from one inhomogeneous patch to another. The 6 specific Szekeres regions, A to F described below, are assembled into different sequences along the light paths, that we call models 1 to 5.

To completely define a Szekeres model, the freedom in the radial coordinate r must be fixed, and five functions of r need to be specified. In this section, all models will be defined by the functions: k, M, S, P and Q. The algorithm used in the calculations can be defined as follows:

1. The radial coordinate is chosen to be the areal radius at the last-scattering instant, t_1: $r' = \Phi(r, t_1)$. For clarity in further use, the prime is omitted.
2. The chosen background model is the ΛCDM model, i.e. a flat Friedmann model with $\Lambda \neq 0$. The background density at the current instant is then given by

$$\rho_b = \Omega_m \times \rho_{cr} = 0.27 \times \frac{3H_0^2}{8\pi G},$$

where the Hubble constant is $H_0 = 72$ km s^{-1} Mpc^{-1}. The cosmological constant, Λ, corresponds to $\Omega_\Lambda = 0.73$.
3. The initial time, t_1, is chosen to be the time of last scattering, and is calculated using (4.26).
4. Six different Szekeres regions are considered here. Let us denote them as regions A, B, C, D, E and F. The functions M, k, Q, P and S in these regions are defined as follows.

- Regions A and B

$$M = M_b + \begin{cases} M_1 r^3 & \text{for } r \leqslant 0.5a, \\ M_2 \exp\left[-12\left(\frac{r-a}{a}\right)^2\right] & \text{for } 0.5a \leqslant r \leqslant 1.5a, \\ M_1(2a-r)^3 & \text{for } 1.5a \leqslant r \leqslant 2a, \\ 0 & \text{for } r \geqslant 2a, \end{cases}$$

where M_b is the mass in the corresponding volume of the homogeneous universe, $M_b = (4\pi G/3c^2)\rho_{LS}r^3$, $\rho_{LS} = (3H_0^2)/(8\pi G)\ (1+z_{LS})^3$, $M_1 = 8M_2 a^{-3}e^{-3/2}$, M_2 is equal to -0.3 kpc and 0.2 kpc for region A and B respectively, and $a = 12$ kpc.

$$k = -\frac{1}{2} \times \begin{cases} k_1 r^2 & \text{for } r \leqslant 0.5b, \\ k_2 \exp\left[-4\left(\frac{r-b}{b}\right)^2\right] & \text{for } 0.5b \leqslant r \leqslant 1.5b, \\ k_1(2b-r)^2 & \text{for } 1.5b \leqslant r \leqslant 2b, \\ 0 & \text{for } r \geqslant 2b, \end{cases}$$

where $k_1 = 4k_2 a^{-2}e^{-1}$, k_2 is equal to -5.15×10^{-6} and 3.5×10^{-6} for regions A and B respectively, and $b = 10.9$ kpc.

$$S = 1, \quad P = 0, \quad Q = Q_1 \ln(1 + Q_2 r) \times \exp(-Q_3 r),$$

where, for regions A and B respectively, Q_1 equals -0.72 and -1.45, Q_2 equals 1 kpc^{-1} and 0.4 kpc^{-1}, and Q_3 equals 0.01 kpc^{-1} and 0.005 kpc^{-1}. With these definitions the mass distribution and the curvature are the same as in Friedmann models, for $r > 24$ kpc.

- Regions C_1 and C_2

  In region C the functions M and k are the same as in region A. The only difference is in the form of functions S, P and Q which are as follows:

$$S = e^{\alpha r}, \quad P = 0, \quad Q = 0,$$

where α is equal to -0.0255 kpc^{-1} and $+0.0255$ kpc^{-1} for regions C_1 and C_2 respectively. Region C_1 is the mirror image of C_2, where the $Z = 0$ surface is the symmetry plane $[Z \stackrel{\text{def}}{=} \Phi \cos \vartheta$ and ϑ is defined by the stereographic projection (2.151)]. The reason why two mirror-similar regions are employed is that in the coordinates used here, the axial geodesics can only be studied for propagation along the $Z < 0$ direction, in which $\vartheta = -\pi$. Along the $Z > 0$ direction we have $\vartheta = 0$, which corresponds to a point at infinity in the stereographic projection. This problem is overcome by matching C_1 with C_2 along the surface of $Z = 0$. When calculating propagation toward the origin, model C_1 is employed, and when calculating propagation away from the origin, model C_2 is employed. In both models light propagates along the $Z < 0$ axis.

- Regions D_1 and D_2

  In region D the functions M and k are the same as in region B. The only difference is in the form of the functions S, P and Q which are of the following form:

$$S = r^{\alpha}, \quad P = 0, \quad Q = 0,$$

where α equal to -0.97 and $+0.97$ for regions D_1 and D_2 respectively. As above, region D comes from matching regions D_1 and D_2 along the $Z = 0$ surface.

- Regions E and F

$$M = M_b + \begin{cases} M_1 r^3 & \text{for } r \leqslant 0.5a, \\ M_2 \exp\left[-6\left(\frac{r-a}{a}\right)^2\right] & \text{for } r \geqslant 0.5a, \end{cases}$$

$$k = \begin{cases} k_1 r^2 & \text{for } r \leqslant 0.5b, \\ k_2 \exp\left[-\left(\frac{r-b}{0.5b}\right)^2\right] & \text{for } r \geqslant 0.5b, \end{cases}$$

$$S = 1, \quad P = 0, \quad Q = Q_1 - 0.22 \ln(1 + Q_2 r) \times \exp(-Q_3 r),$$

where $M_1 = 8a^{-3} M_2 e^{-1.5}$, $k_1 = 4a^{-2} k_2 e^{-1}$. For region E, $M_2 = -0.75$ kpc, $a = 15.23$ kpc, $k_2 = -1.00173 \times 10^{-5}$, $b = 12.95$ kpc, $Q_1 = -0.22$, $Q_2 = 1$ kpc^{-1}, $Q_3 = 0.1$ kpc^{-1}. For region F, $M_2 = 0.9$ kpc, $a = 23.76$ kpc, $k_2 = 7 \times 10^{-6}$, $b = 19.1$ kpc, $Q_1 = -1.4$, $Q_2 = 0.4$ kpc^{-1}, $Q_3 = 0.005$ kpc^{-1}.

5. Light propagation was calculated by solving (3.70)–(3.73) (for models 1 and 3) and (3.93) (for models 2, 4 and 5) simultaneously with the evolution equation (2.146). At each step the null condition, $k_\alpha k^\alpha = 0$, was used to test the

precision of calculations. All equations were solved using the fourth-order Runge–Kutta method.

6. The temperature fluctuations were calculated from (7.3) (for models 1 and 3) and (3.100) (for models 2, 4 and 5). The mean redshift $\bar{z}$ was calculated using the ΛCDM model.

The density distribution at the current instant within each of these regions is presented in Fig. 7.1. Regions A–D are specified in such a way that for $r \geqslant 24$ kpc they become the Friedmann model. Regions E and F tend exponentially to the Friedmann models. However, as seen from their specification (see above) or from Fig. 7.2, at the distance $r \approx 30$ kpc and $r \approx 40$ kpc, respectively, regions E and F become almost Friedmann. Figure 7.2 presents the curvature scalar, $\mathcal{W}^2$, which is defined as

$$\mathcal{W}^2 = \frac{E_{\alpha\beta}E^{\alpha\beta}}{6H^4}, \tag{7.4}$$

where $E_{\alpha\beta}$ is the electric part of the Weyl tensor (2.155) and $H = (1/3)\theta$ is the Hubble parameter [see (2.153)]. As can be seen, in some regions $\mathcal{W}^2 \gg 1$. This feature, apart from nonlinear evolution and non-symmetrical shape, makes the application of the Szekeres model more realistic.

7.1.3 Arrangement of the Swiss-cheese model

The Swiss-cheese models which are employed here are constructed from a Friedmann background and many inhomogeneous Szekeres patches, using six different building blocks – spherical regions A–F. These Szekeres patches are placed so that their boundaries touch wherever a light ray exits one inhomogeneous patch; thus the ray immediately enters another Szekeres inhomogeneity and spends no time in the Friedmann background (except for model 3, where light propagates for a while in the Friedmann region). Five models are constructed, using different sequences of regions A–F; see Fig. 7.3 for a schematical representation of the Swiss-cheese model.

When constructing a Swiss-cheese model, we need to satisfy the junction conditions for matching the particular inhomogeneous patches to the Friedmann background, and also assure the continuity of the null geodesics. The standard junction conditions are that the 3D metric of the surface and its extrinsic curvature, the first and second fundamental forms, must be continuous.

For matching a Szekeres patch to a Friedmann background across a comoving spherical surface, $r =$ constant, the conditions are: that the mass inside the junction surface in the Szekeres patch is equal to the mass that would be inside that surface in the homogeneous background; that the spatial curvature function at the junction surface is the same in both the Szekeres and Friedmann models, $k_{SZ} = k_F r^2$; and the bang time must be continuous across the

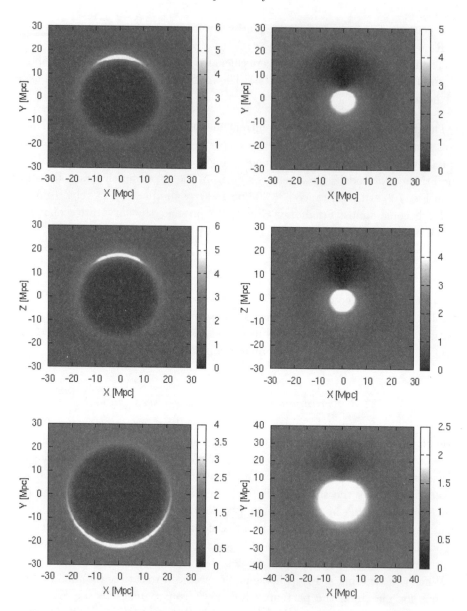

Fig. 7.1. The density distribution, ρ/ρ_b, at the current instant for Szekeres regions A (upper left), B (upper right), C (middle left), D (middle right), E (lower left) and F (lower right). In regions A, B, E and F the dipole axis is aligned with the Y-axis. For these regions the density distribution is presented for the surface $Z = 0$. In regions C and D the dipole is aligned with the Z-axis. For these regions the density distribution is presented for the surface $Y = 0$.

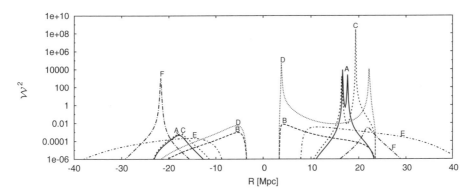

Fig. 7.2. The Weyl curvature scalar $\mathcal{W}^2$ at the present time, as defined by (7.4), along the dipole axis, R, in Szekeres regions A–F. (For regions A, B, E and F, 'R' is the Y-axis, and for regions C and D, 'R' is the X-axis). The 3D shape of these profiles is very similar to the shape of the density distribution – see Fig. 7.1. Since the FLRW models are conformally flat, the Szekeres regions are in some parts far from being even a close (linear) approximation of the Friedmann model.

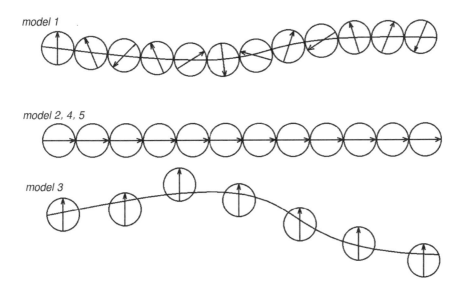

Fig. 7.3. Sketches of the Szekeres Swiss-cheese models. Each sketch shows a light ray (solid curve) passing through a sequence of Szekeres patches (circles). The arrows in the circles show the direction of the Szekeres dipoles – aligned or random. The Szekeres patches are the 'holes' and the 'cheese' is the Friedmann background. It is envisaged that there are many Szekeres patches closely packed together, but only the ones the light passes through are shown.

junction.[2] (We have assumed the same value of Λ for both to make the matching possible.)

The junction of null geodesics requires the continuity of all components of the null vector. However, let us notice that when one Szekeres sphere is matched to another Szekeres sphere it can be rotated around the normal direction. Thus, as discussed in Sec. 3.4, we only need to match up the time component k^0 and the tangential component:

$$k^T = \frac{\Phi}{\mathcal{E}} \sqrt{(k^x)^2 + (k^y)^2}\bigg|_{\text{region 1}} = \frac{\Phi}{\mathcal{E}} \sqrt{(k^x)^2 + (k^y)^2}\bigg|_{\text{region 2}}. \qquad (7.5)$$

The radial component is then given by the null condition, $k_\alpha k^\alpha = 0$.

Five different Szekeres Swiss-cheese models are considered here:

1. Model 1

   Model 1 is constructed from alternately matching regions A and B (A + B + A + B ...) into the Friedmann background. When a light ray exits one Szekeres region, it immediately enters another inhomogeneous patch. Each time, the (x, y) position of the point of entry is randomly selected. In addition k^x and k^y are quasi-randomly selected, i.e.

   $$(k^y)^2 = \gamma \left(k^T \frac{\mathcal{E}}{\Phi}\right)^2, \quad (k^x)^2 = (1 - \gamma) \left(k^T \frac{\mathcal{E}}{\Phi}\right)^2,$$

   where γ is a random value in the range $0 \leqslant \gamma \leqslant 1$. The radial coordinate of the matching point is $r_j = 24$ kpc – the point where the Szekeres region becomes Friedmann. A schematic representation of this model is presented in Fig. 7.3.

2. Model 2

   This model is constructed from alternating regions C and D, but only axial null geodesics are considered, i.e. $k^x = 0$ and $k^y = 0$, $x = y = 0$. The radial component of the matching point is again $r_j = 24$ kpc. See Fig. 7.3 for a schematic representation.

3. Model 3

   The next model consists of regions E and F placed alternately. Null vector components k^x and k^y are chosen in such a way that $10^{-8} \leqslant k^x \leqslant 10^{-4}$ and $10^{-8} \leqslant k^y \leqslant 10^{-4}$, but are otherwise random. As can be noted, this is not in accordance with condition (7.5). In order to maintain the continuity of the tangential component of the null vector the next Szekeres patch must by reoriented with respect to the preceding patch. This however would lead to an overlapping of successive Szekeres regions. To eliminate such an overlap, the Szekeres patches are separated from each other by a distance $\Delta r = 40$

---

[2] It might be surprising that a nonsymmetrical model like the Szekeres one can be joined with the symmetric Friedmann model, but there are other examples of such junctions. For example Bonnor (1976) showed that the Szekeres model can be matched to the Schwarzschild solution.

kpc (which corresponds to a current distance of approximately 40 Mpc). The model is shown schematically in Fig. 7.3.

4. Model 4

   Model 4 is constructed using only C regions, with $r_j = 24$ kpc, and only axial geodesics are considered, i.e. $k^x = k^y = x = y = 0$.

5. Model 5

   The last model is also axially symmetric, $k^x = k^y = x = y = 0$, $r_j = 24$ kpc, but uses only D regions.

In a generic cosmological model, as proved by Sato *et al.* (see the summary by Sato, 1984), voids are not compensated, i.e. the mass within a void is smaller than the mass of the background that would occupy the same space. Such voids expand faster than the background, and once they begin to collide, they flatten against each other. In constructing the Swiss-cheese model illustrated by Fig. 7.3 we assumed that the inhomogeneities are compensated, i.e. properly matched into the Friedmann background and co-expanding with it. We did this in order to isolate one contribution to the anisotropy of the CMB temperature: that from the mass inhomogeneities.

7.1.4 The Rees–Sciama effect

To estimate the temperature fluctuations induced by the light propagation effects, it is assumed that initial temperature distribution is uniform, $(\Delta T/T)_e = 0$. Then temperature fluctuations are calculated using (7.3), and they are plotted against time of propagation in Fig. 7.4. As seen, the final values are small, of amplitude $\Delta T/T \approx 10^{-7}$ (model 3), $\Delta T/T \approx 10^{-6}$ (models 1 and 2), and $\Delta T/T \approx 10^{-5}$ (models 4 and 5). A detailed analysis of how inhomogeneities induce temperature fluctuations is presented in Fig. 7.5 (for clarity, only a small fraction of the time is presented). The left panel of Fig. 7.5 shows the density of regions through which the light propagates in model 3. The right panel presents the temperature fluctuations as measured by an observer situated at the junction point where the model is that of Friedmann. Letters correspond to each inhomogeneous patch (left panel) and temperature fluctuations caused by them (right panel). Clearly, underdense regions induce negative temperature fluctuations, overdense regions induce positive fluctuations.

Apart from estimating the amplitude of the Rees–Sciama effect, it is also important to estimate the angular scale which is the most affected by this effect. Without going into any complicated analysis, we can estimate the power spectrum by employing the following approximation: the correlation between two distant points on the sky is zero – photons which were propagating along two widely separated paths have the temperature fluctuations uncorrelated. Only when the light paths are near to each other are the temperature fluctuations correlated. Thus the simplest estimation of the angular scale of the Rees–Sciama

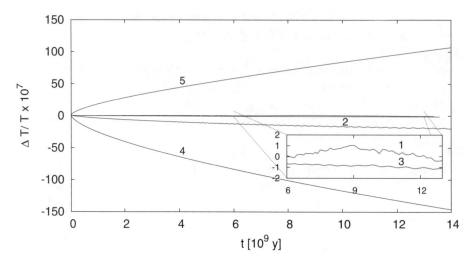

Fig. 7.4. The temperature fluctuations caused by light propagation effects in models 1–5. In models 1–3 light propagates alternatively through underdense and overdense regions. In model 4 light propagates only through regions with $\delta M < 0, \delta k > 0$, and in model 5 only through regions with $\delta M > 0, \delta k < 0$.

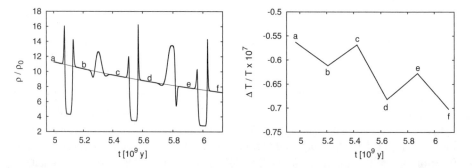

Fig. 7.5. A small part of the light propagation in model 3. The left panel shows the density variation that light 'feels' as it propagates. The black thin dotted line shows the density in the background model. The right panel presents the temperature fluctuations as measured by an observer situated outside the structure in the homogeneous Friedmann region. The letters in the left and right panels label corresponding points along the light path.

effect, as seen from the schematic Fig. 7.6, is the angular size of the Szekeres patch at the last-scattering instant. For the models studied in this section, such approximations lead to an angular scale of $\theta \approx 0.21°$, or alternatively $\ell \approx 850$. If the photons are propagating along neighbouring paths for only half of the age of the Universe (in such a case, as seen from Fig. 7.7, the final temperature fluctuations are two times smaller), then the angular scale is similar, $\theta \approx 0.24°$ ($\ell \approx 750$). Thus, the Rees–Sciama effect of amplitude $\sim 10^{-6}$ contributes to the CMB temperature fluctuations on the angular scale $\theta < 0.25°$ ($\ell > 700$). This angular scale

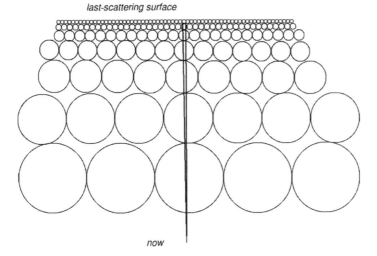

Fig. 7.6. A schematic representation of the Swiss-cheese model. When two photons are propagating along neighbouring paths, the final temperature fluctuations are similar. If the paths are well separated then the final temperature fluctuations are different, and hence not correlated.

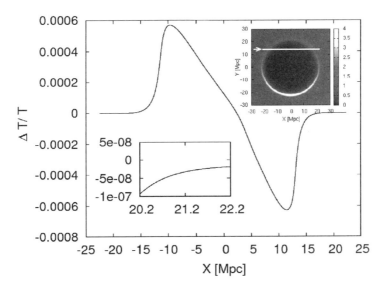

Fig. 7.7. Temperature fluctuation amplitude, as measured by observers at different locations in region E, along the path of a light ray. The ray path is shown as the bright line in the upper inset.

corresponds to the angular scale at which the third peak of the CMB angular power spectrum is observed. At this scale the measured rms temperature fluctuations are of amplitude $\approx 5 \times 10^{-5}$. This is still several times higher than the results obtained within models 4 and 5. In the case of models 1–3 the measured value is more than one order of magnitude larger than the model estimates.[3]

7.1.5 The role of local structures

So far, it has been assumed that each inhomogeneous structure is compensated (i.e. each Szekeres region was matched with the Friedmann background), and that measurements are carried out away from the inhomogeneities, i.e. where the Universe is homogeneous. However in the real Universe there is no place where the cosmic structures in the observer's vicinity are fully compensated and therefore the Universe should not be treated locally as homogeneous. Since all measurements are always local, let us consider what happens if the temperature fluctuations are measured in an uncompensated region. Figure 7.7 presents the temperature fluctuations measured by an observer situated at different places within region E. These results are obtained under the assumption that light from last scattering is propagating through homogeneous regions, and currently reaches an observer in an inhomogeneous structure (region E). The light enters and propagates along the bright line shown in the upper-right inset of Fig. 7.7. The above results show that local structures can significantly contribute to the CMB temperature fluctuations. This indicates that care must be exercised when extracting information from the CMB observations. Although it is highly unlikely that the signal caused by the local structures has a signature of acoustic oscillations, we should be aware that local structures can have some visible impact on observations. Thus it is important to test if local structures can indeed be responsible for the observed correlations of the alignment of dipole, quadrupole and octopole axes of the angular power spectrum of the CMB temperature fluctuations (Schwarz *et al.*, 2004; Vale, 2005; Land and Magueijo, 2005, 2007; Rakić *et al.*, 2006; Rakić and Schwarz, 2007).

The next section will examine how large-scale inhomogeneities can affect the amplitude of low-order multipoles.

7.2 CMB dipole and quadrupole in the DBB model with an off-centre observer

The dipole moment in the CMB anisotropy is the most prominent feature in the observational data. Its value being of order 10^{-3}, it exceeds the quadrupole, of

---

[3] However, within the models studied here, at the current instant all inhomogeneities are of diameters $\sim$50–60 Mpc. Larger inhomogeneities, of diameters $\sim$600 Mpc, can have more significant impact, see e.g. Inoue and Silk (2006, 2007) or Masina and Notari (2009).

order $5 \cdot 10^{-6}$, by more than two orders of magnitude (Smoot *et al.*, 1992; Kogut *et al.*, 1993; Hinshaw *et al.*, 2007).

This dipole is usually considered as resulting from a Doppler effect produced by our motion with respect to the CMB rest frame (Partridge, 1988). A few authors (Gunn, 1988; Paczyński and Piran, 1990; Turner, 1991; Langlois, 1996; Langlois and Piran, 1996) intended however to show that its origin could be in the large-scale features of the Universe. In particular, Paczyński and Piran (1990), using an ad hoc toy model, emphasised the possibility that the dipole might be generated by an entropy gradient in an L–T universe. In the model they studied, they assumed that the Big Bang time was the same for all observers, i.e. $t_B(r) = \text{const.}$

Working in the framework of the DBB class of L–T models described in Sec. 6.3, Schneider and Célérier (1999) showed that the dipole and quadrupole anisotropy, or parts of them, could be viewed as the outcome of a conic Big Bang surface when the observer is located off the symmetry centre. We sum up below their contribution.

7.2.1 Definition of the temperature

The specific entropy S is usually defined as the ratio of the number density of photons to the number density of baryons, i.e.

$$S \equiv \frac{k_B \, n_\gamma(t, r) m_b}{\rho(t, r)}, \tag{7.6}$$

where m_b is the baryon mass and k_B the Boltzmann constant.

In order to decouple the effect of an inhomogeneous entropy distribution (as already studied by Paczyński and Piran, 1990) from that of an inhomogeneous singularity surface, we set $S = \text{const.}$

The observed deviation of the CMB from perfect isotropy being very small, we can assume, as a reasonable approximation, thermodynamical equilibrium for the photons, so at the ultra-relativistic limit for bosons,

$$n_\gamma = a_n T^3, \tag{7.7}$$

T being the radiation temperature and $a_n = 2\zeta(3)k_B^3/[\pi^2(\hbar c)^3]$ (ζ is the Riemann zeta function).

Letting, with no loss of generality, $S = \text{const} = k_B \eta_0$, and taking the present photon to baryon density ratio η_0 to be $10^8/(2.66\Omega_b h_0^2)$, we obtain the following expression for T:

$$T(t, r) = \left(\frac{10^8}{2.66 h_0^2 2\pi G a_n m_b (3t - 5t_B(r))(t - t_B(r))} \right)^{1/3}, \tag{7.8}$$

where h_0 is the Hubble constant in units $100 \text{ km s}^{-1} \text{ Mpc}^{-1}$. Hereafter, for numerical applications, the value $h_0 = 0.75$ will be assumed.

7.2.2 Integration of the null geodesics and determination of the dipole and quadrupole moments

Light travels on the past light cone from an emission 2-surface $(t_{\ell s}, r_{\ell s})$ at last scattering, to the observer at (t_0, r_0). The optical depth of the Universe to Thomson scattering is approximated by a step function. The last-scattering surface is therefore defined, in the local thermodynamical equilibrium approximation discussed above, by its temperature, $T = 4000$ K, as is the now-surface, $T = 2.73$ K, where the observer is located. The equal-temperature surfaces obey (7.8).

Since the value of the entropy function $S(r)$ is assumed to be constant, the constant T curves are monotonically increasing with r, and they asymptotically approach the shell-crossing surface (6.7).

An observer O located at a distance r_0 from the centre sees an axially symmetric universe, with the axis passing through the observer and the centre of symmetry C. The photon path is uniquely defined by the observer's position (t_0, r_0) and the angle α between the direction from which the light ray comes and the direction of the centre.

Figure 7.8 shows the geometry of incoming light rays. If observed at an angle $\alpha \leq \pi/2$, a light ray emitted from point A on the last-scattering surface approaches C to a comoving distance $r_{\min}$, then proceeds toward O. From the opposite direction, the ray follows the geodesic BO.

The system of three differential equations which specifies the null geodesics defining the past light cone of the observer is given by (3.19)–(3.21).

Provided one chooses the affine parameter λ increasing from $\lambda = 0$ at (t_0, r_0) to $\lambda = \lambda_{\ell s}$ at $(t_{\ell s}, r_{\ell s})$ on the last-scattering surface, one has to consider two cases:

- the 'out-case': the observer looks in a direction away from the centre of symmetry $(\alpha > \pi/2)$. The null geodesics are therefore integrated from (t_0, r_0) to $(t_{\ell s}, r_{\ell s})$ with an always increasing r. Equation (3.20) is written with a plus sign.
- the 'in-case': the observer observes a light ray that first approaches the centre of symmetry then moves away from it, before reaching her telescope $(\alpha < \pi/2)$. Equation (3.20) with the minus sign first obtains until $dr/d\lambda = 0$, then changes to a plus sign.

A number of 'in' and 'out' null geodesics, each characterised by a value for α between zero and $\pi/2$, are integrated back in time from the observer at $[t_0, r_0, (k^t)_0, T_0]$ until the temperature, as given by (7.8), reaches $T_{\ell s} = (4/2.7) \times 10^3 T_0$, which defines approximately the last-scattering surface.

At this temperature, the redshift with respect to the observer, as given by (3.14), is $z_{\ell s}^{\text{in,out}}(\alpha)$, which varies somewhat with α and with the 'in' and 'out' direction, about an average $z_{\ell s}^{\text{av}}$. Now, the apparent temperature of the CMB as

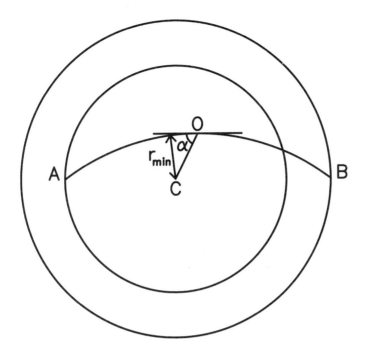

Fig. 7.8. The CMB observed from O with an angle α. Schematic illustration of the trajectory of two CMB light beams received by the observer O and making an angle α with the direction of the symmetry centre C of the universe. The two-dimensional projection of the last-scattering surface seen by O is approximately a circle centred on O and passing through A and B. In general it will intersect many comoving shells of constant r. Points A and B are seen at antipodal points by O, and their comoving shells are solid lines. (Note that OB is not a continuation of AO, since O is the vertex of the light cone.)

measured in the α in-out direction is:

$$T_{\mathrm{CMB}}^{\mathrm{in,out}}(\alpha) = \frac{T_{\ell s}}{1 + z_{\ell s}^{\mathrm{in,out}}(\alpha)} = T_{\mathrm{CMB}}^{\mathrm{av}} \frac{1 + z_{\ell s}^{\mathrm{av}}}{1 + z_{\ell s}^{\mathrm{in,out}}(\alpha)}, \tag{7.9}$$

where the averages for T and z are calculated over the whole sky, and in simplified notation this is

$$\frac{T_{\mathrm{CMB}}}{T^{\mathrm{av}}} = \frac{1 + z^{\mathrm{av}}}{1 + z_{\ell s}}. \tag{7.10}$$

The large-scale CMB temperature inhomogeneities are expanded in spherical harmonics as

$$\frac{T_{\mathrm{CMB}}(\alpha, \varphi)}{T^{\mathrm{av}}} = \sum_{\ell=1}^{\infty} \sum_{m=-\ell}^{+\ell} a_{\ell m} Y_{\ell m}(\alpha, \varphi), \tag{7.11}$$

with

$$a_{\ell m} = \int \frac{T_{\mathrm{CMB}}(\alpha, \varphi)}{T^{\mathrm{av}}} Y_{\ell m}^{*}(\alpha, \varphi) \sin \alpha \, \mathrm{d}\alpha \, \mathrm{d}\varphi. \tag{7.12}$$

The dipole and quadrupole moments are defined as

$$D = (|a_{1-1}|^2 + |a_{10}|^2 + |a_{11}|^2)^{1/2}, \tag{7.13}$$

and

$$Q = (|a_{2-2}|^2 + |a_{2-1}|^2 + |a_{20}|^2 + |a_{21}|^2 + |a_{22}|^2)^{1/2}. \tag{7.14}$$

In the special case we are interested in, the large-scale inhomogeneities depend only on the angle α, so that all the $a_{\ell m}$ with $m \neq 0$ are zero. Therefore the dipole and quadrupole moments reduce to

$$D = a_{10}, \qquad Q = a_{20}. \tag{7.15}$$

It follows that

$$D = (1 + z^{\mathrm{av}}) \int_0^\pi \frac{Y_{10}(\alpha)}{1 + z_{\ell s}(\alpha)} \sin \alpha \, \mathrm{d}\alpha, \tag{7.16}$$

$$Q = (1 + z^{\mathrm{av}}) \int_0^\pi \frac{Y_{20}(\alpha)}{1 + z_{\ell s}(\alpha)} \sin \alpha \, \mathrm{d}\alpha. \tag{7.17}$$

Taking into account (3.14) and the spherical symmetry of the model, one obtains

$$D = \left| \frac{1}{2} \sqrt{\frac{3}{\pi}} k^t_{\mathrm{av}} \left[\int_0^{\frac{\pi}{2}} \frac{\cos \alpha \sin \alpha}{k^t_{\mathrm{in}}(\alpha)} \mathrm{d}\alpha - \int_0^{\frac{\pi}{2}} \frac{\cos \alpha \sin \alpha}{k^t_{\mathrm{out}}(\alpha)} \mathrm{d}\alpha \right] \right|, \tag{7.18}$$

$$Q = \frac{1}{4} \sqrt{\frac{5}{\pi}} k^t_{\mathrm{av}} \left[\int_0^{\frac{\pi}{2}} \frac{(3 \cos^2 \alpha - 1) \sin \alpha}{k^t_{\mathrm{in}}(\alpha)} \mathrm{d}\alpha + \int_0^{\frac{\pi}{2}} \frac{(3 \cos^2 \alpha - 1) \sin \alpha}{k^t_{\mathrm{out}}(\alpha)} \mathrm{d}\alpha \right].$$
$$\tag{7.19}$$

7.2.3 Results and discussion

A class of DBB models exhibiting a Big Bang function of the form

$$t_B(r) = br^n, \qquad b > 0, \ n > 0 \tag{7.20}$$

was identified by Célérier and Schneider (1998) as solving the horizon problem. In a later paper (Schneider and Célérier, 1999) the subclass chosen to be investigated was for simplicity,

$$t_B(r) = br \quad \text{with} \quad 1/R_H > b > 0. \tag{7.21}$$

This conic surface corresponds to perturbations with low spatial frequencies, $k < 1/R_H$, R_H being the horizon radius.

After having numerically integrated back in time from the observer a number of 'in' and 'out' null geodesics, each characterised by a value for α between zero and $\pi/2$, up to the last-scattering surface, as explained above, the dipole and quadrupole moments D and Q were calculated, according to (7.18) and (7.19).

Table 7.1. *Best-fit values of r_0 and b for reproducing the CMB dipole and quadrupole values using a DBB model.*

r_0	b	D	Q
0.02	2×10^{-7}	1.61×10^{-3}	5.27×10^{-5}
0.03	9×10^{-8}	1.11×10^{-3}	3.70×10^{-5}
0.04	7×10^{-8}	1.15×10^{-3}	3.99×10^{-5}
0.05	6×10^{-8}	1.23×10^{-3}	4.57×10^{-5}
0.06	5×10^{-8}	1.23×10^{-3}	5.33×10^{-5}
0.07	4×10^{-8}	1.15×10^{-3}	5.79×10^{-5}

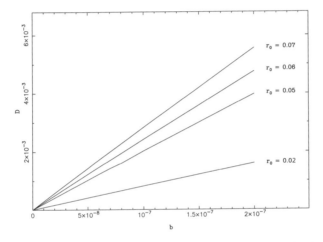

Fig. 7.9. The dipole amplitude in a DBB model, as a function of b, for various values of r_0.

The best-fit values of the doublets (r_0, b) that lead to D and Q close to the observed values, $D \sim 10^{-3}$, $Q \sim 10^{-5}$, are given in Table 7.1. The variations of the dipole and quadrupole as functions of b for various values of r_0 are plotted in Figs. 7.9 and 7.10.

Using a toy model, chosen within the class of DBB models, it has been shown that values for the model parameters, i.e. the location of the observer in space-time and the increasing slope of the bang function, can be found that closely reproduce the observed dipole and quadrupole moments in the CMB anisotropy. This provides a possible cosmological interpretation of the dipole (or part of it, as it is obvious that there is probably a Doppler component due to the local motion of the Galaxy with respect to the CMB rest frame).

As has been stressed by other authors (Paczyński and Piran, 1990; Turner, 1991; Langlois, 1996; Langlois and Piran, 1996), there are various observational methods to discriminate between a local and a cosmological origin for the dipole. If, from future analyses of observational data, part of the dipole appears to be

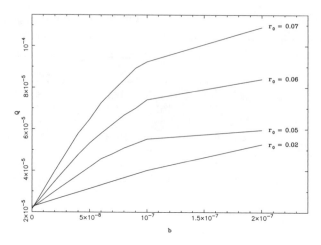

Fig. 7.10. The quadrupole amplitude in a DBB model, as a function of b, for various values of r_0.

non-Doppler, other work, connected in particular with multipole moments of higher order, would be needed to discriminate between the various cosmological candidate interpretations.

Provided the observations allow such a discrimination to be carried out, any unexplained part of the dipole compatible with the above model might set correlated bounds upon the location of the observer and the slope of the Big Bang function, adding therefore new constraints on the parameters of this type of L–T model.

A step toward such a goal was made in Alnes and Amarzguioui (2006), with the use of another particular class of L–T models, which we describe in the next section.

7.3 CMB anisotropies in the Alnes *et al.* model with an off-centre observer

We return to the notation of Sec. 3.1.3. Alnes and Amarzguioui (2006) concentrated on the study of the anisotropies arising in the CMB temperature if the observer is shifted to an off-centre location, analysing two specific realisations of the model of Alnes *et al.* (2006). One of these realisations is the model best fitting the SN Ia data, while the second one is slightly different. To disregard any intrinsic anisotropies on the last-scattering surface, they assumed the temperature to be isotropic there.

Following the line of thought already detailed in Sec. 7.2, they noted first that the CMB temperature as measured today by the observer is

$$T(\xi) = \frac{T_*}{1 + z(\xi)},$$
(7.22)

where T_* is the temperature at last scattering and ξ the angle denoted α in Sec. 7.2. The average temperature $\hat{T}$ given by the Planck spectrum as seen by the observer is therefore

$$\hat{T} \equiv \frac{1}{4\pi} \int d\Omega\, T(\xi) = \frac{T_*}{2} \int_0^\pi d\xi \frac{\sin \xi}{1 + z(\xi)}. \tag{7.23}$$

According to Smoot *et al.* (1992), this temperature is $\hat{T} = 2.725$ K. An average redshift for the last-scattering surface can therefore be defined from (7.22) and (7.23) that reads

$$1 + z_* \equiv \frac{T_*}{\hat{T}} = 2 \left[\int_0^\pi d\xi \frac{\sin \xi}{1 + z(\xi)} \right]^{-1}. \tag{7.24}$$

This redshift is obtained by integrating the set of differential equations for $(t, r, \vartheta, p$ and $z)$ given in Sec. 3.1.3 with the initial conditions $[t_0, r_0, 0, p_0$ and $z_0 = z(\lambda = 0) = 0]$. The relative temperature variation reads thus

$$\theta \equiv \frac{\Delta T}{\hat{T}} = \frac{T(\xi) - \hat{T}}{\hat{T}} = \frac{z_* - z(\xi)}{1 + z(\xi)}. \tag{7.25}$$

Then, as in Sec. 7.2, the contributions at different angular scales were considered by decomposing the temperature variation into spherical harmonics. The magnitude of the observed dipole (of order 10^{-3}) was used to put constraints on how far away from the centre the observer may be shifted, since, as shown by Schneider and Célérier (1999), a location farther off-centre usually yields a larger dipole. Recombination was assumed to occur at $t = 0$ and the present time was defined as the time when the redshift of the photons emitted at $t = 0$ reaches $z_* \simeq 1100$.

The geodesic equations were solved for two positions of the observer, 20 Mpc and 200 Mpc from the centre, in both models. Then, the coefficients $a_{\ell 0}$ for the dipole ($\ell = 1$), the quadrupole ($\ell = 2$) and the octopole ($\ell = 3$) were calculated as functions of the observer position.

As expected from Schneider and Célérier (1999), the induced multipoles become larger the farther away from the centre the observer is located. The distance of the observer from the centre is severely constrained by the dipole value alone. For the induced dipole to be $\lesssim 10^{-3}$, the physical distance of the observer from the centre should satisfy $d_{\mathrm{obs}} \lesssim 15$ Mpc. When compared to the size of the underdensity in the models, which is around 1500 Mpc, the authors argued that this amounts to a rather strong violation of the Copernican 'principle', but weaker than in the central observer case. We disagree with this evaluation. If the observer were at a spot with exactly average density, this too would be highly unlikely. Where must one be in a Universe full of inhomogeneities in order not to violate the Copernican principle? In the real Universe, there is nowhere that is not in an inhomogeneity. Although finding ourselves exactly at the middle of an inhomogeneity is somewhat non-Copernican, as long as that inhomogeneity is just one of many it is not very strongly non-Copernican. We suggest 5% off-centre is as Copernican as 50% off-centre.

However, the most striking result is that the quadrupole and the octopole are too small as compared to the dipole. If the induced dipole value is assumed to be around the observed 10^{-3}, the induced quadrupole becomes smaller than 10^{-7} and the induced octopole is less than 10^{-9}, their behaviour being similar in both models. This is far less than their measured values $\sim 10^{-5}$. It is thus clear that these models cannot explain the observed alignment of the low-ℓ multipoles in the CMB spectrum.

Even if part of the dipole might be assigned to a nonvanishing peculiar velocity of the observer with respect to the CMB photon flow, which can be fixed to any arbitrary value, the problem would remain, since the multipoles would be expected to keep the hierarchical scaling they exhibit in this particular kind of model. (Using a simplified Newtonian framework, it is easy to show that for these models the dipole scales linearly, the quadrupole quadratically and the octopole cubically with the observer location.) Therefore, even if one can manage to fit the correct values of the dipole and quadrupole, the octopole would still remain too small.

As a conclusion, the authors claimed that their models are not ruled out by these results, even if they are unable to explain the low multipole anomaly. We want however to stress that the models studied in this work are only special cases of L–T models from which it might be premature to draw general conclusions. Nevertheless, the features thus exhibited are striking, and, given that there is plenty of structure in our local Universe, these effects must play some role.

8
Conclusions

A central aim of this book is to show that there is much still to be learned about the evolution and effects of exact nonlinear inhomogeneities in Einstein's theory, and that this is highly relevant to the real Universe.

The Universe as we observe it is very inhomogeneous. Among its structures there are groups and clusters of galaxies, large cosmic voids and very large elongated structures such as filaments and walls. In cosmology, however, the homogeneous and isotropic models of the Robertson–Walker class have been used almost exclusively, and in these, structure formation is described by an approximate perturbation theory. This works well as long as the perturbations remain small, but cannot be applied once perturbations become large and evolution becomes nonlinear. This is where the methods of inhomogeneous cosmology must come into play. These methods can be employed not only to study the evolution of cosmic structures, but also to investigate the formation and evolution of black holes, as well as studying the geometry and dynamics of the Universe.

Whatever the successes of the Concordance model, based on an FLRW metric plus perturbation theory, structure evolution does sooner or later become nonlinear and non-Newtonian, and our understanding of present-day observations will be incomplete without the methods of inhomogeneous cosmology. The phenomena of fully relativistic inhomogeneous evolution must occur and cannot be ignored.

This book presents the application of inhomogeneous exact solutions of the Einstein equations in such areas as the evolution of galactic black holes and cosmic structures, and the impact of inhomogeneities on light propagation which allows us to solve the horizon and dark energy problems. The following is a brief summary of its contents and results.

1. **Evolution of galactic black holes**

   When studying galactic black holes, it is commonly assumed that these objects can be described using the Schwarzschild or Kerr metric. However the Schwarzschild or Kerr black holes do not evolve, they just exist unchanging for all time. To describe the formation and evolution of black holes, nonstationary models must be employed. In Sec. 4.2 we have described the nonlinear evolution of an initial density perturbation at recombination into a galaxy with a central black hole at the present day, using the spherically symmetric L–T model. Assuming the presence of a central black hole today, a density profile based on observational data for M87 was taken for the galaxy exterior to the horizon, and a black hole model with mass $3 \cdot 10^9 M_\odot$ was smoothly joined on as the interior.

Since no observational data exist (and, presumably, will not exist for a long time) concerning the interior of the horizon, two distinct forms of this central black hole – both L–T models – were considered: firstly a condensation that collapses to a singularity, and secondly a full Schwarzschild–Kruskal–Szekeres type wormhole with nonzero density. It has been demonstrated that within these two cases it is possible to obtain a realistic model of a galactic black hole. In the collapse case, the black hole forms only half a billion years before the present. In the wormhole case, the wormhole is only open for a fraction of a second, having minimum mass of a couple of $M_\odot$, but has already accreted a quarter of a million $M_\odot$ by recombination. Both models grow to $3 \cdot 10^9 M_\odot$ by today.

In view of the paucity of data, our approach was the first exploratory step into an uncharted territory rather than an actual model to be compared with observations.

2. **Evolution of cosmic structures**

The evolution of cosmic structures is usually described either in the linear perturbation formalism or in N-body simulations. However, these methods face some problems and limitations. The linear approach is not valid once the density contrast becomes large. Moreover, in the linear regime one cannot study the impact of shape on evolution since, in this approximation, the shape of a structure does not affect its evolution (Barrow and Silk, 1981).

The limitations of the N-body simulations are: the use of Newtonian mechanics, the assumption of a uniform Hubble expansion, and the finite number of particles (Joyce and Marcos, 2007).

Therefore investigations based on exact solutions are mandatory.

We have used exact relativistic models such as the L–T, Lemaître and Szekeres models to reproduce the formation of structures in the Universe.

In Sec. 2.1.5 methods of generating the L–T models that evolve between given initial and final profiles of either density or velocity were presented. These methods have been used in Sec. 4.3 to describe the formation and evolution of the Abell cluster A2199. In Sections 4.4 and 4.5 we have presented models of cosmic voids. Section 4.6 shows the evolution of pairs consisting of a galaxy supercluster and a void, and of triple structures consisting of a galaxy supercluster, a void and a wall.

We have found that velocity perturbations are more efficient at generating structures than density perturbations. The gravitational instability paradigm, which focusses on analysing Fourier modes rather than individual structures, is probably not telling us the whole story. This point is strengthened by the fact that an initial condensation can evolve into a void, and vice-versa.

It has also been found that the shape and environment of cosmic structures have an impact on evolution. Namely, small voids among large overdense regions do not evolve as fast as large isolated voids do. This is because the expansion of the space is faster inside large voids than inside smaller voids. Moreover, this higher expansion rate inside large voids leads to the formation of large and elongated structures such as walls and filaments which emerge at the edges of these large voids. These elongated structures evolve much faster than compact overdense regions (the density contrast within these structures increases faster).

The angular distribution of the CMB temperature must be known at much finer scales before any models of structure formation can be truly tested against these observations. The current best resolution of $0.1°$ is to be compared with the approximate angular sizes of the portions of matter on the CMB sky that will go into a galaxy cluster, $\sim 0.1°$, and into a single galaxy, $\sim (10^{-3}–10^{-2})°$, see Sec. 4.1.1.

One unsatisfactory feature of the L–T and Szekeres models used here is that once matter starts collapsing, it will keep collapsing to a singularity. In real objects, rotation, pressure gradients and non-perfect fluid behaviour can prevent collapse so that the object virialises. More general exact cosmological solutions will be needed if we are to create highly accurate models that remain valid as collapse sets in.

3. **The horizon problem**
An important application of exact inhomogeneous models is to study the early stages of the Universe and its large-scale homogenisation. The central problem here is whether the Universe started homogeneous (or very close to homogeneous) or the currently observed large-scale homogeneity has developed during its evolution. Currently, this second opinion is in favour among cosmologists, and the mechanism which leads to homogenisation is believed to be inflation. However, inflation does not really solve this problem, but merely explains that regions of the Universe which now enter our horizon were causally connected before the inflation epoch. The inflationary hypothesis only postpones the occurrence of the horizon problem to a later time, after which the observer will be once more confronted with it. This raises the question, why are we living at a time when inflation still solves the horizon problem? Moreover, inflation replaces one problem of initial conditions with another, i.e. the specific initial conditions of the Universe (homogeneity) are replaced by mechanisms which are specially designed to drive inflation and whose 'natural' production has become a major topic in the literature.

In Chapter 6 we gave a resolution of the horizon problem in the framework of a large class of L–T Delayed Big Bang models, which solve it permanently, unlike the inflation paradigm. The geometry of these L–T models allows our past light cone to bend sufficiently to never cross the Big Bang surface and therefore solves the causality problem at last scattering for ever.

4. **Cosmological observations in the inhomogeneous universe**

Another aspect of this book was to employ inhomogeneous models to analyse cosmological observations such as supernovae, the cosmic microwave background, the baryon acoustic oscillation scale, etc., which are usually only analysed against homogeneous models.

In Chapter 5 we demonstrated that the L–T models are able in principle to reproduce the observed dimming of the distant supernovae with no need for a cosmological constant or, in any case, that an effect of inhomogeneities on the determination of the cosmological parameters might not be negligible. However, the luminosity data do not fully specify an L–T model, leaving one arbitrary function undetermined, even in the case of non-spherically symmetric models, e.g. Swiss-cheese models where the holes in a Friedmannian background are represented by L–T solutions.

We have stressed that the right way of dealing with this issue is to try to reproduce *not* an *accelerating* Universe but the Hubble diagram of the SN Ia. Actually, an inhomogeneous model can mimic the magnitude–redshift relation of the Concordance model without exhibiting an accelerating expansion. This is the reason why no-go theorems do not apply here (see Sec. 5.1.4).

The CMB temperature fluctuations have been studied in the framework of L–T and Szekeres models (Chapter 7). The observed values of the dipole, the quadrupole and the octopole have been recovered by shifting our location away from the centre of particular inhomogeneous models. However, even if future analyses of observational data show that part of the dipole appears to be of cosmological origin, more work connected with multipole moments of higher order would be needed to discriminate between various cosmological candidate interpretations.

The effect of light propagation in a lumpy Universe on the CMB temperature fluctuations has been studied in Sec. 7.1. The Rees–Sciama effect was analysed by propagating light rays through Swiss-cheese models where the holes are Szekeres regions matched into a Friedmann background. When each underdense region is followed by an overdense region the final fluctuations are small, of order 10^{-7}. But when light is propagated only through underdense or overdense regions, the final values are by one order of magnitude higher.

Our results show that the L–T and Szekeres models are very useful tools for this kind of investigation. However, for the parameters of the models to be properly fitted to the results of observations, the observational data would have to be re-interpreted against the background of these inhomogeneous geometries. This calls for a thorough revision of the whole existing body of data and literature – a process that will take a long time.

The traditional FLRW cosmological models are the simplest solutions of Einstein's equations with nonzero expansion. They were used at the beginning of relativistic cosmology to account for the redshift of the galaxies observed in the 1920s by Hubble and Humason. However, they make use of relativity in a most simplistic way, which is doomed to fail when trying to account for an accurate description of our Universe at all cosmological scales. In the further development of astrophysical cosmology based on these models, the main input from relativity was the assumption that the Universe was hot and dense in the past, and then cooled as it expanded. Everything else is thermodynamics and particle physics. It is usually claimed that the light element abundances confirming the standard picture of nucleosynthesis, the nearly isotropic CMB black body power spectrum and the Hubble law are three observational pillars which sustain the robustness of FLRW models. However, even in a Friedmannian framework, observations are not sufficiently precise to define the evolution uniquely, so there is certainly enough room to consider more general models. It is often claimed that the high isotropy of the CMB radiation (with maximal temperature anisotropies of $\Delta T / T \approx 2.93 \times 10^{-5}$ at the angular scale of about $0.9°$; see Hinshaw *et al.*, 2009), combined with the Copernican principle, *proves* that our Universe is homogeneous and exhibits therefore a Robertson–Walker geometry (see Ehlers *et al.*, 1968; and Stoeger *et al.*, 1995). Such statements often ignore the caveats stated in Stoeger *et al.* (1995), as well as the smallness of the effect of matter inhomogeneities on the temperature of propagating radiation. Existing estimates (see Chapter 7) show that the interaction is weak, and no temperature anisotropies larger than 10^{-5} should ever have been expected.

By considering more general (L–T, Lemaître and Szekeres) cosmological models, one does not in any way deny the confirmed successes of the FLRW cosmology. The more general models should fill in many aspects that cannot be captured in the FLRW geometries, like structure formation. The FLRW models still remain valid as a rough first approximation to a more detailed description.

We do not claim that we have found solutions to all problems and mysteries of modern cosmology. However, we do wish to register a protest against the certainty with which unverified and often currently untestable theories have been advanced to 'explain' some observed phenomena, and the light-heartedness with which inhomogeneous alternatives are dismissed. Apparently, this happens in order to remain within the most elementary cosmological model. Such attitudes are rather non-scientific in their spirit. Before we decide that we do have an explanation of a set of observations, we should carefully consider all possible theoretical descriptions, and, true to the time-honoured tradition, choose the one that is logically simplest and does without introducing untestable assumptions. To concentrate on a few examples:

1. An early example is the inflationary prediction that $\Omega_0 = 1$. In fact, inflation predicts Ω_0 is very close to unity, and it does not increase the probability that Ω_0 is exactly 1, which remains vanishingly small. While calculations in a flat FLRW model give good approximations for many purposes, there are qualitative differences between flat and non-flat models. Indeed there are problems for inflation in FLRW models with positive spatial curvature (Ellis *et al.*, 2002). Despite this, for a long time many advocates stated it is exactly 1.

2. Cold dark matter was introduced in order to allow perturbations to begin growing before recombination, because otherwise structures would not form soon enough to agree with observations. Its equation of state has been adjusted to improve the properties of the present-day matter distribution. Although astronomers had long mooted the existence of dark matter to account for galaxy rotation curves etc., neither the equation of state nor the amount were at all like current proposals.

3. Extra-galactic type Ia supernova measurements indicate a decrease of apparent luminosities, relative to expectations in zero pressure, zero Λ Friedmann models. Since these larger than expected luminosity distances don't fit our once-preferred model, and since homogeneity is not to be questioned, we force the calculated dimming to agree with our favourite simple cosmological model, by declaring that 70% of the current energy density of the Universe is 'dark energy' whose properties nobody can explain.

4. In fact we boldly declare that we know how the scale factor evolved, while freely admitting we don't know the physical nature of more than 95% of the content of the Universe. (Currently there are several tens of different explanations for dark energy and dark matter, starting from brane worlds and ending with aether-like scalar fields or a Chaplygin gas.) Although within such a model one obtains concordance, i.e. the right CMB power spectrum, an explanation of the dimming of supernovae and the calculated age of the Universe comparable to the age of the oldest stars, it is actually a collection of phenomenological proposals that lack a coherent physical basis. Given these uncertainties, it is far too soon to declare that the Concordance or ΛCDM model is a true image of our Universe. It should be noted that there already exist explanations of the CMB power spectrum and other cosmological observations based on general relativity that do not invoke any entities unknown from the laboratory, but simply take into account that inhomogeneities in the matter distribution do exist, and do influence the collection and interpretation of observational results (Godłowski *et al.*, 2004; Alnes *et al.*, 2006; Wiltshire, 2007; Alexander *et al.*, 2007; García-Bellido and Haugbølle, 2008a; Sarkar, 2008).

The history of science should teach us some caution. It happened before that ad-hoc fixes proposed in order to solve an isolated problem developed into elaborate research paradigms. Well-known examples are the aether hypothesis and the steady-state theory, the first proposed to provide a natural reference frame for electromagnetic waves, the second to solve the 'age problem' of the Universe. In both cases, the solution of the original problem was found by taking the good old theory to farther limits than before, which led to new interesting developments. We propose to do the same now with the apparent 'acceleration of expansion'. We showed in this book that very promising attempts at explaining this phenomenon within standard relativity, without invoking 'dark energy', already exist in the literature.

A curious tendency can be observed in modern cosmology. At the end of 19th century most scientists firmly believed that Newtonian mechanics can be successfully employed to describe the world. However, the enormous intellectual and technological progress of the 20th century, which made very subtle and sophisticated experiments possible, proved something opposite. Nowadays most scientists have no doubts that there is plenty to discover about the true nature of the Universe. However, at the beginning of the 21st century many tend to look for *new physics* rather than to seek an explanation within the current paradigm. The main point of this book is a demonstration that a solid part of relativistic cosmology can be done by exact methods of relativity, without applying approximations, heuristic methods of dubious mathematical foundation or by referring to *new physics*. Additionally, whatever new physics emerges, the methods and results of inhomogeneous cosmology still need to be taken into account. The L–T model has been used many times over the years for a wide variety of applications, yet it continues to generate interesting results. The Szekeres model on the other hand has only recently been used for modelling, and has a much greater potential for a very broad range of interesting applications and results. The diversity of possible models and uses is essentially unknown, and we can be sure that more useful applications will emerge. Especially with the Szekeres models, there is enormous scope for investigating the formation and evolution of complex structures.

This book shows that relativity has a lot more to offer to cosmology than just the standard homogeneous models of the Robertson–Walker class, and it also demonstrates that inhomogeneous metrics can quite easily generate realistic models of cosmic structures and their nonlinear formation and evolution.

To end on an optimistic note, let us observe that in recent years appreciation of inhomogeneous models in the astrophysical community has increased in a dramatic way. In 1994, when one of us (Krasiński, 1997) completed the review of inhomogeneous cosmological models, it included just *all* papers in which *any* aspect of such models had been discussed until then. Up to that time, papers on such topics had been published predominantly in physical and mathematical journals. By today, there is a regular stream of new publications discussing various astrophysical and observational applications of the L–T model, they are often published in astronomical journals, and the preferred medium for their pre-publication is the astro-ph archive, not gr-qc as before. It would be rather impossible to capture them in a review of reasonable size, and any such review would instantly become outdated. The Szekeres model is slowly making its way into the field. Thus, it seems that we are witnessing the birth of a new-style theoretical astrophysics that will not be slavishly bound to just one class of cosmological models, but will allow general model-fitting as a legitimate activity.

References

Alexander, S., Biswas, T., Notari, A. and Vaid D. (2007). Local void vs dark energy: confrontation with WMAP and type Ia supernovae, arXiv:0712.0370. [**172, 214**]

Alnes, H. and Amarzguioui, M. (2006). CMB anisotropies seen by an off-center observer in a spherically symmetric inhomogeneous universe, *Phys. Rev.* **D74**, 103520. [**55, 58, 59, 156, 157, 159, 172, 206**]

Alnes, H. and Amarzguioui, M. (2007). Supernova Hubble diagram for off-center observers in a spherically symmetric inhomogeneous universe, *Phys. Rev.* **D75**, 023506. [**58, 59, 156, 157, 159, 172**]

Alnes, H., Amarzguioui, M. and Grøn, Ø. (2006). Inhomogeneous alternative to dark energy?, *Phys. Rev.* **D73**, 083519. [**156, 157, 159, 160, 161, 172, 206, 214**]

Apostolopoulos, P. S., Brouzakis, N., Tetradis, N. and Tzavara E. (2006). Cosmological acceleration and gravitational collapse, *J. Cosmol. Astropart. Phys.* **06(2006)**, 009. [**146, 147**]

Araújo, M. E., Arcuri, R. C., Bedran, M. L., de Freitas, L. R. and Stoeger, W. R. (2001). Integrating Einstein field equations in observational coordinates with cosmological data functions: nonflat Friedmann–Lemaître–Robertson–Walker cases *Astrophys. J.* **549**, 716–20. [**181**]

Araújo, M. E. and Stoeger, W. R. (2009). The angular-diameter distance maximum and its redshift as constraints on $\Lambda \neq 0$ FLRW models, *Mon. Not. R. Astron. Soc.* **394**, 438. [**172**]

Araújo, M. E., Stoeger, W. R., Arcuri, R. C. and Bedran, M. L. (2008). Solving Einstein field equations in observational coordinates with cosmological data functions: spherically symmetric universes with cosmological constant, *Phys. Rev.* **D78**, 063513. [**181**]

Arnau, J. V., Fullana, M. J., Monreal, L. and Saez, D. (1993). On the microwave background anisotropies produced by nonlinear voids, *Astrophys. J.* **402**, 359. [**189**]

Arnau, J. V., Fullana, M. J., Saez, D. (1994). Great Attractor-like structures and large-scale anisotropy, *Mon. Not. R. Astron. Soc.* **268**, L17. [**190**]

Baade, W. (1956). The period–luminosity relation of the Cepheids, *Publ. Astron. Soc. Pac.* **68**, 5. [**139**]

Barrett, R. K. and Clarkson, C. A. (2000). Undermining the cosmological principle: almost isotropic observations in inhomogeneous cosmologies, *Class. Quant. Grav.* **17**, 5047. [**148**]

Barrow, J. D. and Silk, J. (1981). The growth of anisotropic structures in a Friedmann Universe, *Astrophys. J.* **250**, 432. [**210**]

Barrow, J. D. and Stein-Schabes, J. (1984). Inhomogeneous cosmologies with cosmological constant, *Phys. Lett.* **A103**, 315. [**47**]

Begelman, M. C. (2003). Evidence for black holes, *Science* **300**, 1898. [**94**]

Bene, G., Czinner, V. and Vasúth, M. (2006). Accelerating expansion of the universe may be caused by inhomogeneities, *Mod. Phys. Lett.* **A21**, 1117. [**147**]

Bennett, C. L., Halpern, M., Hinshaw, G., Jarosik, N., Kogut, A., Limon, M., Meyer, S. S., Page, L., Spergel, D. N., Tucker, G. S., Wollack, E., Wright, E. L., Barnes, C., Greason, M. R., Hill, R. S., Komatsu, E., Nolta, M. R., Odegard, N., Peiris, H. V., Verde, L. and Weiland, J. L. (2003). First-Year Wilkinson Microwave Anisotropy Probe (WMAP) Observations: Preliminary Maps and Basic Results, *Astrophys. J. Suppl. Ser.* **148**, 1. [**109**]

Biswas, T. and Notari, A. (2008). 'Swiss-Cheese' Inhomogeneous Cosmology and the Dark Energy Problem, *J. Cosmol. Astropart. Phys.* **06(2008)**, 021. [**164, 167, 169, 172, 173**]

Biswas, T., Mansouri, R. and Notari A. (2007). Nonlinear Structure Formation and 'Apparent' Acceleration: an Investigation, *J. Cosmol. Astropart. Phys.* **12(2007)**, 017. [**172**]

Bolejko, K. (2006a). Structure formation in the quasispherical Szekeres model, *Phys. Rev.* **D73**, 123508. **[94, 126]**

Bolejko, K. (2006b). Radiation in the process of the formation of voids, *Mon. Not. R. Astron. Soc.* **370**, 924. **[116]**

Bolejko, K. (2007). Evolution of cosmic structures in different environments in the quasispherical Szekeres model, *Phys. Rev.* **D75**, 043508. **[94, 126]**

Bolejko, K. (2008). Supernovae Ia observations in the Lemaître–Tolman model, *PMC Physics* **A2**, 1. **[149, 172]**

Bolejko, K. and Andersson, L. (2008). Apparent and average acceleration of the Universe, *J. Cosmol. Astropart. Phys.* **10(2008)**, 003. **[147]**

Bolejko, K. and Hellaby, C. (2008). The Great Attractor and the Shapley Concentration, *Gen. Rel. Grav.* **40**, 1771. **[33]**

Bolejko, K., Krasiński, A. and Hellaby, C. (2005) Formation of voids in the Universe within the Lemaître–Tolman model. *Mon. Not. R. Astron. Soc.* **362**, 213. **[30, 125]**

Bolejko, K. and Lasky, P. D. (2008) Pressure gradients, shell crossing singularities and acoustic oscillations – application to inhomogeneous cosmological models. *Mon. Not. R. Astron. Soc.*, **391**, L59. **[114]**

Bolejko, K. and Wyithe, J. S. B. (2009) Testing the Copernican Principle Via Cosmological Observations. *J. Cosmol. Astropart. Phys.* **02(2009)**, 020. **[149, 172, 173]**

Bonamente, M., Joy, M. K., LaRoque, S. J., Carlstrom, J. E., Reese, E. D. and Dawson, K. S. (2006). Determination of the Cosmic Distance Scale from Sunyaev-Zel'dovich Effect and Chandra X-Ray Measurements of High-Redshift Galaxy Clusters, *Astrophys. J.* **647** 25. **[173]**

Bondi, H. (1947). Spherically symmetrical models in general relativity *Mon. Not. R. Astron. Soc.* **107**, 410; reprinted, with historical comments, in *Gen. Rel. Grav.* **31**, 1777 (1999). **[143]**

Bonnor, W. B. (1976). Non-radiative solutions of Einstein's equations for dust, *Commun. Math. Phys.* **51**, 191. **[47, 196]**

Bonnor, W. B. (1985). An open recollapsing cosmological model with $\Lambda = 0$, *Mon. Not. R. Astron. Soc.* **217**, 597. **[68]**

Brouzakis, N. and Tetradis, N. (2008). Analytical estimate of the effect of spherical inhomogeneities on luminosity distance and redshift, *Phys. Lett.* **B665**, 344. **[166]**

Brouzakis, N., Tetradis, N. and Tzavara, E. (2007). The effect of large scale inhomogeneities on the luminosity distance, *J. Cosmol. Astropart. Phys.* **02(2007)**, 013. **[164, 166, 169, 172]**

Brouzakis, N., Tetradis, N. and Tzavara, E. (2008). Light propagation and large-scale inhomogeneities, *J. Cosmol. Astropart. Phys.* **04(2008)**, 008. **[164, 166, 169, 172]**

Bruna, L. and Girbau, J. (1999a). Linearization stability of the Einstein equation for Robertson-Walker models. I, *J. Math. Phys.* **40**, 5117. **[3]**

Bruna, L. and Girbau, J. (1999b). Linearization stability of the Einstein equation for Robertson-Walker models. II, *J. Math. Phys.* **40**, 5131. **[3]**

Bruna, L. and Girbau, J. (2005). Laplacian in the hyperbolic space H_n and linearization stability of the Einstein equation for Robertson–Walker models, *J. Math. Phys.* **46**, 072501. **[3]**

Buchert, T. (2000). On average properties of inhomogeneous fluids in general relativity: dust cosmologies, *Gen. Rel. Grav.* **32**, 105. **[2]**

Buchert, T. (2001). On average properties of inhomogeneous fluids in general relativity: perfect fluid cosmologies, *Gen. Rel. Grav.* **33**, 1381. **[2]**

Buchert, T. (2008). Dark energy from structure – a status report, *Gen. Rel. Grav.* **40**, 467. **[2]**

Buchert, T. and Carfora, M. (2002). Regional averaging and scaling in relativistic cosmology, *Class. Quant. Grav.* **19**, 6109. **[2]**

Buchert, T. and Carfora, M. (2003). Cosmological parameters are dressed, *Phys. Rev. Lett.*. **90**, 031101. **[2]**

Caldwell, R. R. and Stebbins, A. (2008). A test of the Copernican principle, *Phys. Rev. Lett.* **100**, 191302. **[173]**

Capozziello, S., Cardone, V. F. and Troisi, A. (2006). Dark energy and dark matter as curvature effects, *J. Cosmol. Astropart. Phys.* **08(2006)**, 001. **[137]**

Carroll, S. M., Duvvuri, V., Trodden, M. and Turner, M. S. (2004). Is cosmic speed-up due to new gravitational physics?, *Phys. Rev.* **D 70**, 043528. **[137]**

Carroll, S. M., Press, W. H. and Turner, E. L. (1992). The cosmological constant, *Ann. Rev. Astron. Astrophys.* **30**, 499. **[140]**

Célérier, M. N. (2000a). Do we really see a cosmological constant in the supernovae data?, *Astron. Astrophys.* **353**, 63. **[55, 138, 140, 145, 172, 177, 179, 180]**

Célérier, M. N. (2000b). Models of universe with a delayed big-bang singularity III. Solving the horizon problem for an off-center observer, *Astron. Astrophys.* **362**, 840. **[56, 184, 188]**

Célérier, M. N. (2007a). The accelerated expansion of the Universe challenged by an effect of inhomogeneities: a review, *New Adv. Phys.* **1**, 29. **[138, 172]**

Célérier, M. N. (2007b). Inhomogeneities in the Universe and the fitting problem, to be published in the *Proceedings of the XIXth Rencontres de Blois, Matter and energy in the Universe*, Blois, France, May 2007; arXiv: 0706.1029 [astro-ph]. **[7]**

Célérier, M. N., Bolejko, K., Krasiński, A. and Hellaby, C. (2009). A central underdensity is not mandatory to explain away dark energy with a Lemaître – Tolman model. arxiv 0906.0905. **[173]**

Célérier, M. N. and Schneider, J. (1998). A solution to the horizon problem: a delayed Big Bang singularity, *Phys. Lett.* **A249**, 37. **[145, 184, 186, 188, 204]**

Célérier, M. N. and Szekeres, P. (2002). Timelike and null focusing singularities in spherical symmetry: a solution to the cosmological horizon problem and a challenge to the cosmic censorship hypothesis, *Phys. Rev.* **D 65**, 123516. **[184, 185]**

Christodoulou, D. (1984). Violation of cosmic censorship in the gravitational collapse of a dust cloud, *Commun. Math. Phys.* **93**, 171. **[65]**

Chung, D. J. H. and Romano, A. E. (2006). Mapping luminosity–redshift relationship to Lemaitre–Tolman–Bondi cosmology, *Phys Rev.* **D74**, 103507. **[172, 177]**

Clifton, T., Ferreira, P. G. and Land, K. (2008). Living in a void: testing the Copernican principle with distant supernovae, *Phys. Rev. Lett.* **101**, 131302. **[172, 173]**

Crocce, M. and Scoccimarro, R. (2006). Renormalized cosmological perturbation theory, *Phys. Rev.* **D73**, 063519. **[3]**

Cyburt, R. H. (2004). Primordial nucleosynthesis for the new cosmology: determining uncertainties and examining concordance, *Phys. Rev.* **D70**, 023505. **[155]**

Dąbrowski, M. P. and Hendry, M. A. (1998). The Hubble diagram of type ia supernovae in nonuniform pressure universes, *Astrophys. J.* **498**, 67. **[138, 148, 172]**

de Souza, M. M. (1985). Hidden symmetries of Szekeres quasi-spherical solutions, *Revista Brasileira de Física* **15**, 379. **[52]**

Di Matteo, T., Allen, S. W., Fabian, A. C., Wilson, A. S. and Young, A. J. (2003). Accretion onto the supermassive black hole in M87, *Astrophys. J.* **582**, 133–140. **[100]**

Doran, M. and Lilley, M. (2002). The location of cosmic microwave background peaks in a universe with dark energy, *Mon. Not. R. Astron. Soc.* **330**, 965. **[154]**

Dunsby, P. K. S. (1997). A fully covariant description of cosmic microwave background anisotropies, *Class. Quant. Grav.* **14**, 3391. **[93]**

Dwivedi, I. H. and Joshi, P. S. (1992). Cosmic censorship violation in non-self-similar Tolman–Bondi models, *Class. Quant. Grav.* **9**, L69. **[65]**

Eardley, D. M. and Smarr, L. (1979). Time functions in numerical relativity: marginally bound dust collapse, *Phys. Rev.* **D19**, 2239. **[65]**

Ehlers, J., Geren, P. and Sachs, R. K. (1968). Isotropic solutions of the Einstein–Liouville equations, *J. Math. Phys.* **9**, 1344–9. **[189, 213]**

Einstein, A. and Straus, E. G. (1945). The influence of the expansion of space on the gravitation fields surrounding the individual stars, *Rev. Mod. Phys.* **17**, 120 (see also erratum: Einstein, A. and Straus, E. G. (1946). Corrections and Additional Remarks to our Paper: The Influence of the Expansion of Space on the Gravitation Fields Surrounding the Individual Stars, *Rev. Mod. Phys.* **18**, 148–49). **[163]**

Eisenstein, D. J. and Hu, W. (1998). Baryonic features in the matter transfer function, *Astrophys. J.* **496**, 605. [154]

Eisenstein, D. J., Zehavi, I., Hogg, D. W., Scoccimarro, R., Blanton, M. R., Nichol, R. C., Scranton, R., Seo, H.–J., Tegmark, M., Zheng, Z., Anderson, S. F., Annis, J., Bahcall, N., Brinkmann, J., Burles, S., Castander, F. J., Connolly, A., Csabai, I., Doi, M., Fukugita, M., Frieman, J. A., Glazebrook, K., Gunn, J. E., Hendry, J. S., Hennessy, G., Ivezić, Z., Kent, S., Knapp, G. R., Lin, H., Loh, Y.–S., Lupton, R. H., Margon, B., McKay, T. A., Meiksin, A., Munn, J. A., Pope, A., Richmond, M. W., Schlegel, D., Schneider, D. P., Shimasaku, K., Stoughton, C., Strauss, M. A., SubbaRao, M., Szalay, A. S., Szapudi, I., Tucker, D. L., Yanny, B. and York, D. G. (2005). Detection of the Baryon Acoustic Peak in the Large-Scale Correlation Function of SDSS Luminous Red Galaxies, *Astrophys. J.* **633**, 560. [149, 158]

Ellis, G. F. R. (1971). Relativistic cosmology, in *Proceedings of the International School of Physics 'Enrico Fermi', Course 47: General Relativity and Cosmology*, ed. R. K. Sachs. Academic Press, New York and London, pp. 104 – 182; reprinted, with historical comments, in *Gen. Rel. Grav.* **41**, 581 (2009). [57, 112, 139]

Ellis, G. F. R. (2007). On the definition of distance in general relativity: I. M. H. Etherington, *Gen. Rel. Grav.* **39**, 1047–52. [139]

Ellis, G. F. R. (2008). Patchy solutions, *Nature* **452**, 158. [2, 189]

Ellis, G. F. R., Nel, S. D., Maartens, R., Stoeger, W. R. and Whitman, A. P. (1985). Ideal observational cosmology, *Phys. Rep.* **124**, 315. [159, 181]

Ellis, G. F. R. and Stoeger, W. (1987). The 'fitting problem' in cosmology, *Class. Quant. Grav.* **4**, 1697. [137, 169]

Ellis, G. F. R., Stoeger, W., McEwan, P. and Dunsby, P. (2002). Dynamics of inflationary universes with positive spatial curvature. *Gen. Rel. Grav.* **34**, 1445–9. [214]

Enqvist, K. (2008). Lemaitre–Tolman–Bondi model and accelerating expansion, *Gen. Rel. Grav.* **40**, 451. [172]

Enqvist, K. and Mattsson, T. (2007). The effect of inhomogeneous expansion on the supernova observations, *J. Cosmol. Astropart. Phys.* **02(2007)**, 019. [147, 170, 172]

Etherington, I. M. H. (1933). On the definition of distance in general relativity, *Phil. Mag. VII* **15**, 761–73 [reprinted, with historical comments, in *Gen. Rel. Grav.* **39**, 1055–67 (2007)]. [139]

Fabricant, D., Lecar, M. and Gorenstein, P. (1980). X-ray measurements of the mass of M87, *Astrophys. J.* **241**, 552. [98, 99]

Flanagan, E. E. (2005). Can superhorizon perturbations drive the acceleration of the Universe?, *Phys. Rev.* **D71**, 103521. [146]

Ford, H. C., Harms, R. J., Tsvetanov, Z. I., Hartig, G. F., Dressel, L. L., Kriss, G. A., Bohlin, R. C., Davidsen, A. F., Margon, B. and Kochhar, A. K. (1994). Narrowband HST images of M87: evidence for a disk of ionized gas around a massive black hole, *Astrophys. J.* **435**, L27. [97]

Freedman, W. L., Madore, B. F., Gibson, B. K., Ferrarese, L., Kelson, D. D., Sakai, S., Mould, J. R., Kennicutt, R. C. Jr., Ford, H. C., Graham, J. A., Huchra, J. P., Hughes, S. M. G., Illingworth, G. D., Macri, L. M. and Stetson, P. B. (2001). Final results from the Hubble Space Telescope Key Project to measure the Hubble constant, *Astrophys. J.* **553**, 47. [171]

Frith, W. J., Busswell, G. S., Fong, R., Metcalfe, N. and Shanks, T. (2003). The local hole in the galaxy distribution: evidence from 2MASS, *Mon. Not. R. Astron. Soc.* **345**, 1049. [172]

Fullana, M. J., Saez, D. and Arnau, J. V (1994). On the microwave background anisotropy produced by Great Attractor-like structures, *Astrophys. J. Suppl. Ser.* **94**, 1. [190]

García-Bellido, J. and Haugbølle, T. (2008a). Confronting Lemaitre–Tolman–Bondi models with observational cosmology, *J. Cosmol. Astropart. Phys.* **04(2008)**, 003 . [155, 161, 172, 214]

García-Bellido, J. and Haugbølle, T. (2008b). Looking the void in the eyes – the kinematic Sunyaev–Zeldovich effect in Lemaître–Tolman–Bondi models, *J. Cosmol. Astropart. Phys.* **09(2008)**, 016. [163, 172]

Geshnizjani, G. and Brandenberger, R. (2002). Back reaction and the local cosmological expansion rate, *Phys. Rev.* **D66**, 123507. **[147]**

Godłowski, W., Stelmach, J. and Szydłowski, M. (2004). Can the Stephani model be an alternative to FRW accelerating models?, *Class. Quant. Grav.* **21**, 3953. **[148, 172, 214]**

Goodman, J. (1995). Geocentrism reexamined, *Phys. Rev.* **D52**, 1821. **[173]**

Gorini, V., Grillo, G. and Pelizza, M. (1989). Cosmic censorship and Tolman–Bondi spacetimes, *Phys. Lett.* **A135**, 154. **[65]**

Gunn, J. E. (1988). Deviations from pure Hubble flow: a review, in *The extragalactic distance scale*, eds. S. van der Bergh and C. Pritchet, ASP Conf. Series **4**, p. 344. **[201]**

Haager, G. (1997). Two-component dust in spherically symmetric motion, *Class. Quant. Grav.* **14**, 2219. **[125]**

Harms, R. J., Ford, H. C., Tsvetanov, Z. I., Hartig, G. F., Dressel, L. L., Kriss, G. A., Bohlin, R., Davidsen, A. F., Margon, B. and Kochhar, A. K. (1994). HST FOS spectroscopy of M87: evidence for a disk of ionized gas around a massive black hole, *Astrophys. J.* **435**, L35. **[97]**

Hellaby, C. (1987). A Kruskal-like model with finite density, *Class. Quant. Grav.* **4**, 635. **[13, 60, 63, 66, 94, 95]**

Hellaby, C. (1988). Volume matching in Tolman models, *Gen. Rel. Grav.* **20**, 1203. **[137]**

Hellaby, C. (1996a). The nonsimultaneous nature of the Schwarzschild $R = 0$ singularity, *J. Math. Phys.* **37**, 2892. **[13, 71]**

Hellaby, C. (1996b). The null and KS limits of the Szekeres model, *Class. Quant. Grav.* **13**, 2537. **[44]**

Hellaby, C. (2001). Multicolour observations, inhomogeneity and evolution, *Astron. Astrophys.*, **372**, 357–63. **[182]**

Hellaby, C. (2006). The mass of the cosmos. *Mon. Not. R. Astron. Soc.* **370**, 239–44. **[172, 176, 182]**

Hellaby, C. and Alfedeel, A.H.A. (2009). Solving the observer metric, *Phys. Rev.* **D79**, 043501. **[181]**

Hellaby, C. and Krasiński, A. (2002). You cannot get through Szekeres wormholes: regularity, topology and causality in quasispherical Szekeres models, *Phys. Rev.* **D66**, 084011. **[42, 50, 51, 74, 83–86, 95]**

Hellaby, C. and Krasiński, A. (2006). Alternative methods of describing structure formation in the Lemaître-Tolman model, *Phys. Rev.* **D73**, 023518. **[13, 22, 33]**

Hellaby, C. and Krasiński, A. (2008). Physical and geometrical interpretation of the $\varepsilon \leq 0$ Szekeres models, *Phys. Rev.* **D77**, 023529. **[42, 44, 45]**

Hellaby, C. and Lake, K. (1985). Shell crossings and the Tolman model, *Astrophys. J.* **290**, 381; erratum *Astrophys. J.* **300**, 461 (1986). **[13, 15, 18, 62, 64, 68, 148, 185, 187]**

Hellaby, C. and Lake, K. (1988). The singularity of Eardley, Smarr and Christodoulou, Preprint 88/7, Department of Applied Mathematics, University of Cape Town. **[65]**

Hinshaw, G., Nolta, M. R., Benett, C. L., Bean, R., Doré, O., Greason, M. R., Halpern, M., Hill, R. S., Jarosik, N., Kogut, A., Komatsu, E., Limon, M., Odegard, N., Meyer, S. S., Page, L., Peiris, H. V., Spergel, D. N., Tucker, G. S., Verde, L., Weiland, J. L., Wollack, E. and Wright, E. L. (2007). Three-year Wilkinson microwave anisotropy probe (WMAP) observations: temperature analysis, *Astrophys. J. Suppl.* **170**, 288. **[201]**

Hinshaw, G., Weiland, J. L., Hill, R. S., Odegard, N., Larson, D., Bennett, C. L., Dunkley, J., Gold, B., Greason, M. R., Jarosik, N., Komatsu, E., Nolta, M. R., Page, L., Spergel, D. N., Wollack, E., Halpern, M., Kogut, A., Limon, M., Meyer, S. S., Tucker, G. S. and Wright, E. L. (2009). Five-year Wilkinson microwave anisotropy probe (WMAP) observations: data processing, sky maps, and basic results, *Astrophys. J. Suppl. Ser.* **180**, 225. **[189, 213]**

Hirata, C. M. and Seljak, U. (2005). Can superhorizon cosmological perturbations explain the acceleration of the universe?, *Phys. Rev.* **D72**, 083501. **[146, 147]**

Hoyle, F. and Vogeley, M. S. (2004). Voids in the two-degree field galaxy redshift survey, *Astrophys. J.* **607**, 751. **[109, 116, 120, 122, 124]**

Humphreys, N. P., Maartens, R. and Matravers, D. R. (1997). Anisotropic observations in universes with nonlinear inhomogeneity, *Astrophys. J.* **477**, 47. **[143, 159]**

Humphreys, N.P., Maartens, R. and Matravers, D.R. (1998). Regular spherical dust spacetimes, arXiv:gr-qc/9804023. **[14]**

Iguchi, H., Nakamura, T. and Nakao, K. (2002). Is dark energy the only solution to the apparent acceleration of the present Universe?, *Prog. Theor. Phys.* **108**, 809. **[172, 174]**

Inoue, K. T. and Silk, J. (2006). Local voids as the origin of large-angle cosmic microwave background anomalies. I., *Astrophys. J.* **648**, 23. **[189, 200]**

Inoue, K. T. and Silk, J. (2007). Local voids as the origin of large-angle cosmic microwave background anomalies: the effect of a cosmological constant, *Astrophys. J.* **664**, 650. **[189, 200]**

Ishibashi, A. and Wald, R. M. (2006). Can the acceleration of our universe be explained by the effects of inhomogeneities?, *Class. Quant. Grav.* **23**, 235. **[147]**

Joshi, P. S. (1993). *Global Aspects in Gravitation and Cosmology.* Clarendon Press, Oxford. **[65]**

Joshi, P. S. and Dwivedi, I. H. (1993). Naked singularities in spherically symmetric inhomogeneous Tolman–Bondi dust cloud collapse, *Phys. Rev.* **D47**, 5357. **[65]**

Joyce, M. and Marcos, B. (2007a). Quantification of discreteness effects in cosmological N-body simulations: initial conditions, *Phys. Rev.* **D75**, 063516. **[210]**

Joyce, M. and Marcos, B. (2007b). Quantification of discreteness effects in cosmological N-body simulations. II. Evolution up to shell crossing, *Phys. Rev.* **D76**, 103505 **[210]**

Kai, T., Kozaki, H., Nakao, K., Nambu, Y. and Yoo, C. M. (2007). Can inhomogeneities accelerate the cosmic volume expansion?, *Prog. Theor. Phys.* **117**, 229. **[147, 164]**

Kantowski, R. (1969). Corrections in the luminosity–redshift relations of the homogeneous Friedmann models, *Astrophys. J.* **155**, 89. **[163]**

Kantowski, R. and Sachs, R. K. (1966). Some spatially homogeneous anisotropic relativistic cosmological models, *J. Math. Phys.* **7**, 443. **[43]**

Kogut, A., Lineweaver, C., Smoot, G. F., Bennett, C. L., Banday, A., Boggess, N. W., Cheng, E. S., De Amici, G., Fixsen, D. J., Hinshaw, G., Jackson, P. D., Janssen, M., Keegstra, P., Loewenstein, K., Lubin, P., Mather, J. C., Tenorio, L., Weiss, R., Wilkinson, D. T. and Wright, E. L. (1993). Dipole anisotropy in the *COBE* differential microwave radiometers first-year sky maps, *Astrophys. J.* **419**, 1. **[201]**

Krasiński, A. (1997). *Inhomogeneous Cosmological Models.* Cambridge University Press. **[3, 7, 11, 25, 94, 215]**

Krasiński, A. (2008). Geometry and topology of the quasiplane Szekeres model, *Phys. Rev.* **D78**, 064038. **[44]**

Krasiński, A. and Hellaby, C. (2002). Structure formation in the Lemaître–Tolman model, *Phys. Rev.* **D65**, 023501. **[16, 18, 19, 22, 92, 101, 111]**

Krasiński, A. and Hellaby, C. (2004a). More examples of structure formation in the Lemaître–Tolman model, *Phys. Rev.* **D69**, 023502. **[16, 20, 93, 94, 101, 106, 171]**

Krasiński, A. and Hellaby, C. (2004b). Formation of a galaxy with a central black hole in the Lemaître–Tolman model, *Phys. Rev.* **D69**, 043502. **[60, 85, 94]**

Krasiński, A., Hellaby, C., Célérier, M. N. and Bolejko, K. (2009). Imitating accelerated expansion of the Universe by matter inhomogeneities – corrections of some misunderstandings. arXiv:0903.4070. **[14]**

Kristian, J. and Sachs, R. K. (1966). Observations in cosmology, *Astrophys. J.* **143**, 379. **[181]**

Kruskal, M. D. (1960). Maximal extension of Schwarzschild metric, *Phys. Rev.* **119**, 1743. **[13]**

Kurki-Suonio, H. and Liang, E. (1992). Relation of redshift surveys to matter distribution in spherically symmetric dust universes, *Astrophys. J.* **390**, 5. **[142]**

Land, K. and Magueijo, J. (2005). Examination of evidence for a preferred axis in the cosmic radiation anisotropy, *Phys. Rev. Lett.* **95**, 071301. **[200]**

Land, K. and Magueijo, J. (2007). The Axis of Evil revisited, *Mon. Not. R. Astron. Soc.* **378**, 153. **[200]**

Langlois, D. (1996). Cosmic microwave background dipole induced by double inflation, *Phys. Rev.* **D54**, 2447. **[201, 205]**

Langlois, D. and Piran, T. (1996). Cosmic microwave background dipole from an entropy gradient, *Phys. Rev.* **D53**, 2908. **[201, 205]**

Leith, B. M., Ng, S. C. C., Wiltshire, D. L. (2008). Gravitational energy as dark energy: concordance of cosmological tests. *Astrophys. J.* **672**, L91. **[2]**

Lemaître, G. (1933). L'Univers en expansion [The expanding Universe], *Ann. Soc. Sci. Bruxelles* **A53**, 51; English translation, with historical comments: *Gen. Rel. Grav.* **29**, 637 (1997). **[4, 11, 13, 40, 41]**

Lemos, J. P. S. (1991). On naked singularities in self-similar Tolman–Bondi spacetimes, *Phys. Lett.* **A158**, 279. **[65]**

Li, N. and Schwarz, D. J. (2007). Onset of cosmological backreaction, *Phys. Rev.* **D76**, 083011. **[3]**

Li, N. and Schwarz, D. J. (2008). Scale dependence of cosmological backreaction, *Phys. Rev.* **D78**, 083531. **[3]**

Losic, B. and Unruh, W.G. (2008). Cosmological perturbation theory in slow-roll spacetimes, *Phys. Rev. Lett.* **101**, 111101. **[3]**

Lu, T. H-C. and Hellaby, C. (2007). Obtaining the spacetime metric from cosmological observations, *Class. Quant. Grav.* **24**, 4107–31. **[176, 181, 182]**

Maartens, R., Humphreys, N.P., Matravers, D.R. and Stoeger, W.R. (1996). Inhomogeneous universes in observational coordinates, *Class. Quant. Grav.* **13**, 253-64; Errata (1996) *Class. Quant. Grav.*, **13**, 1689-90. **[181]**

Macchetto, F., Marconi, A., Axon, D. J., Capetti, A., Sparks, W. and Crane, P. (1997). The supermassive black hole of M87 and the kinematics of its associated gaseous disk, *Astrophys. J.* **489**, 579–600. **[97, 100]**

Marra, V., Kolb, E. W. and Matarrese, S. (2008). Light-cone averages in a Swiss-cheese universe, *Phys. Rev.* **D77**, 023003. **[164, 169]**

Marra, V., Kolb, E. W., Matarrese, S. and Riotto, A. (2007). Cosmological observables in a Swiss-cheese universe, *Phys. Rev.* **D76**, 123004. **[164, 167, 170, 172]**

Masina, I. and Notari, A. (2009). The cold spot as a large void: Rees–Sciama effect on CMB power spectrum and bispectrum, *J. Cosmol. Astropart. Phys.* **02(2009)**, 019. **[190, 200]**

Matarrese, S. (1996). Non-Linear Evolution of Cosmological Perturbations, in: *Proceedings of an Advanced Summer School Held at Laredo*, ed. E. Martinez-Gonzalez and J. L. Sanz, Springer-Verlag, Berlin, Lecture Notes in Physics **470**, p. 131. **[3]**

McClure, M. L. and Hellaby, C. (2008). The metric of the cosmos II: accuracy, stability, and consistency, *Phys. Rev. D* **78**, 044005. **[176, 181, 182]**

Melia, F. and Falcke, H. (2001). The supermassive black hole at the galactic center, *Ann. Rev. Astron. Astrophys.* **39**, 309-52. **[94]**

Mena, F. C. and Tavakol, R. (1999). Evolution of the density contrast in inhomogeneous dust models, *Class. Quant. Grav.*, **16**, 435. **[127]**

Misner, C. W. and Sharp, D. H. (1964). Relativistic equations for adiabatic, spherically symmetric gravitational collapse, *Phys. Rev.* **B136**, 571. **[4, 41]**

Moffat, J. W. (2006a). Late-time inhomogeneity and acceleration without dark energy, *J. Cosmol. Astropart. Phys.* **05(2006)**, 001. **[146, 147]**

Moffat, J. W. (2006b). Late-time inhomogeneity and the acceleration of the Universe, in *Albert Einstein Century International Conference*, ed. J. M. Alimi and A. Füfza. AIP Melville, New York. This article is available only on-line at http://proceedings.aip.org/proceedings/confproceed/861.jsp, p. 987. **[147]**

Mustapha, N., Bassett, B. A. C. C., Hellaby, C. and Ellis, G. F. R. (1998). The distortion of the area distance–redshift relation in inhomogeneous isotropic universes, *Class. Quant. Grav.* **15**, 2363. **[142]**

Mustapha, N. and Hellaby, C. (2001). Clumps into voids, *Gen. Rel. Grav.* **33**, 455. **[14, 108, 111]**

Mustapha, N., Hellaby, C. and Ellis, G. F. R. (1997). Large scale inhomogeneity versus source evolution: can we distinguish them observationally?, *Mon. Not. R. Astron. Soc.* **292**, 817. **[138, 171, 182]**

Navarro, J. F., Frenk, C. S. and White, S. D. M. (1995). Simulations of X-ray clusters, *Mon. Not. R. Astron. Soc.* **275**, 720. **[105]**

Newman, R. P. A. C. (1986). Strengths of naked singularities in Tolman–Bondi spacetimes, *Class. Quant. Grav.* **3**, 527. **[65]**

Newman, R. P. A. C. and Joshi, P. S. (1988). Constraints on the structure of naked singularities in classical general relativity, *Ann. Phys.* **182**, 112. **[65]**

Nilsson, U. S., Uggla, C., Wainwright, J. and Lim, W. C. (1999). An almost isotropic cosmic microwave temperature does not imply an almost isotropic universe, *Astrophys. J.* **521**, L1 **[189]**

Nolan, B. C. and Debnath, U. (2007). Is the shell-focusing singularity of Szekeres space-time visible? *Phys. Rev.* **D76**, 104046. **[75]**

Notari, A. (2006). Late time failure of Friedmann equation. *Mod. Phys. Lett.* **A21**, 2997. **[3]**

Nottale, L. (1982a). Perturbation of the magnitude–redshift relation in an inhomogeneous relativistic model: the redshift equations, *Astron. Astrophys.* **110**, 9. **[163]**

Nottale, L. (1982b). Perturbation of the magnitude–redshift relation in an inhomogeneous relativistic model: II. Correction to the Hubble law behind clusters, *Astron. Astrophys.* **114**, 261. **[163]**

Nottale, L. (1983). Perturbation of the magnitude–redshift relation in an inhomogeneous relativistic model: III. Redshift effect intrinsic to clusters of galaxies, *Astron. Astrophys.* **118**, 85. **[163]**

Nottale, L. (1993). *Fractal space-time and microphysics: towards a theory of scale relativity.* World Scientific. **[138]**

Nottale, L. (1996). Scale relativity and fractal space-time: applications to quantum physics, cosmology and chaotic systems, *Chaos, Solitons & Fractals* **7**, 877. **[138]**

Novikov, I. D. (1964). R- i T-oblasti v prostranstve-vremeni so sfericheski-simetrichnym prostranstvom [R- and T-regions in a spacetime with a spherically symmetric space], *Soobshcheniya GAISh* **132**, 3; English translation, with historical comments: *Gen. Rel. Grav.* **33**, 2255 (2001). **[13]**

Paczyński, B. and Piran, T. (1990). A dipole moment of the microwave background as a cosmological effect, *Astrophys. J.* **364**, 341. **[201, 205]**

Padmanabhan, T. (1996). *Cosmology and Astrophysics Through Problems.* Cambridge University Press. **[93]**

Panek, M. (1992). Cosmic background radiation anisotropies from cosmic structures – models based on the Tolman solution, *Astrophys. J.* **388**, 225. **[189]**

Pang, T. (1997). *An introduction to computational physics.* Cambridge University Press. **[109]**

Partovi, H. H. and Mashhoon, B. (1984). Toward verification of large-scale homogeneity in cosmology, *Astrophys. J.* **276**, 4. **[142, 144]**

Partridge, R. B. (1988). The angular distribution of the cosmic background radiation, *Rep. Progr. Phys.* **51**, 647. **[201]**

Pascual-Sánchez, J. F. (1999). Cosmic acceleration: inhomogeneity versus vacuum energy, *Mod. Phys. Lett.* **A14**, 1539. **[138, 148]**

Peebles, P. J. E. (1980). *The large-scale structure of the Universe.* Princeton University Press. **[109, 119]**

Penrose, R. (1966). General relativistic energy flux and elementary optics, in *Perspectives in geometry and relativity: essays in honour of Vaclav Hlavaty*, ed. B. Hoffman, Indiana University Press, Bloomington, pp. 259–74. **[139]**

Perlmutter, S., Aldering, G., Goldhaber, G., Knop, R. A., Nugent, P., Castro, P. G., Deustua, S., Fabbro, S., Goobar, A., Groom, D. E., Hook, I. M., Kim, A. G., Kim, M. Y., Lee, L. C., Nunes, N. J., Pain, R., Pennypacker, C. R., Quimby, R., Lidman, C., Ellis, R.S., Irwin, M., McMahon, R. G., Ruiz-Lapuente, P. , Walton, N., Schaefer, B., Boyle, B. J., Filippenko, A. V., Matheson, T., Fruchter, A. S., Panagia, N., Newberg, H. J. M. and Couch, W. J. (1999). Measurements of omega and lambda from 42 high-redshift supernovae. *Astrophys. J.* **517**, 565. **[137, 145]**

Plebański, J. and Krasiński, A. (2006). *An Introduction to General Relativity and Cosmology.* Cambridge University Press. **[2, 11, 15, 42–45, 50, 52, 55, 57, 65, 69, 77, 79, 84–86, 94, 139]**

Podurets, M. A. (1964). Ob odnoy forme uravneniy Eynshteyna dlya sfericheski simmetrichnogo dvizheniya sploshnoy sredy [On one form of Einstein's equations for spherically symmetrical motion of a continuous medium], *Astron. Zh.* **41**, 28; English translation: *Sov. Astron. A. J.* **8**, 19 (1964). [41]

Press, W. H., Flannery, B. P., Teukolsky, S. A. and Vetterling, W. T. (1986). *Numerical Recipes. The Art of Scientific Computing.* Cambridge University Press. [109]

Rakić, A., Räsänen, S. and Schwarz, D. J. (2006). The microwave sky and the local Rees–Sciama effect, *Mon. Not. R. Astron. Soc.* **369**, L27. **Note:** This reference is in the online-only section of volume 369, see the link: www.blackwell-synergy.com/doi/abs/10.1111/j.1745-3933.2006.00167.x [190, 200]

Rakić, A. and Schwarz, D. J. (2007). Correlating anomalies of the microwave sky, *Phys. Rev.* **D75**, 103002. [200]

Räsänen, S. (2004). Backreaction in the Lemaître – Tolman – Bondi model, *J. Cosmol. Astropart. Phys.* **11(2004)**, 010. [147]

Räsänen, S. (2006a). Cosmological acceleration from structure formation, *Int. J. Mod. Phys.* D **15**, 2141. [2, 147]

Räsänen, S. (2006b). Accelerated expansion from structure formation, *J. Cosmol. Astropart. Phys.* **11(2006)**, 003. [2, 147]

Rees, M. J. and Sciama, D. W. (1968). Larger scale density inhomogeneities in the Universe, *Nature* **217**, 511. [92]

Riess, A. G., Filippenko, A. V., Challis, P., Clocchiatti, A., Diercks, A., Garnavich, P. M., Gilliland, R. L., Hogan, C. J., Jha, S., Krishner, R. P., Leibundgut, B., Phillips, M. M., Reiss, D., Schmidt, B. P., Schommer, R. A., Smith, R. C., Spyromilio, J., Stubbs, C., Suntzeff, N. B. and Tonry, J. (1998). Observational evidence from supernovae for an accelerating Universe and a cosmological constant, *Astron. J.* **116**, 1009. [137]

Riess, A. G., Strolger, L. G., Tonry, J., Casertano, S., Ferguson, H. C., Mobasher, B., Challis, P., Filippenko, A. V., Jha, S., Li, W., Chornock, R., Kirshner, R. P., Leibundgut, B., Dickinson, M., Livio, M., Giavalisco, M., Steidel, C. C., Benítez, T. and Tsvetanov, Z. (2004). Type Ia supernova discoveries at $z > 1$ from the Hubble Space Telescope: evidence for past deceleration and constraints on dark energy evolution, *Astrophys. J.* **607**, 665. [149, 160, 171]

Riess, A. G., Strolger, L. G., Casertano, S., Ferguson, H. C., Mobasher, B., Gold, B., Challis, P. J., Filippenko, A. V., Jha, S., Li, W., Tonry, J., Foley, R., Kirshner, R. P., Dickinson, M., MacDonald, E., Eisenstein, D., Livio, M., Younger, J., Xu, C., Dahlén, T. and Stern, D. (2007). New Hubble Space Telescope discoveries of type Ia supernovae at $z \geq 1$: narrowing constraints on the early behavior of dark energy, *Astrophys. J* **659**, 98–121. [149]

Romano, A. E. (2007a). Lemaitre–Tolman–Bondi universes as alternatives to dark energy: does positive averaged acceleration imply positive cosmic acceleration?, *Phys. Rev.* **D75**, 043509. [147]

Romano, A. E. (2007b). Redshift spherical shell energy in isotropic universes, *Phys. Rev.* **D76**, 103525. [178]

Sachs, R. K. (1961). Gravitational waves in general relativity VI. The outgoing radiation condition, *Proc. R. Soc.* **A264**, 309. [165]

Sachs, R. K. and Wolfe, A. M. (1967). Perturbations of a cosmological model and angular variations of the microwave background, *Astrophys. J.* **147**, 73; reprinted, with historical comments, in *Gen. Rel. Grav.* **39**, 1944 (2007). [8]

Saez, D., Arnau, J. V. and Fullana, M. J. (1993). The imprints of the Great Attractor and the Virgo cluster on the microwave background, *Mon. Not. R. Astron. Soc.* **263**, 681 [189]

Sarkar, S. (2008). Is the evidence for dark energy secure?, *Gen. Rel. Grav.* **40**, 269. [3, 150, 214]

Sato, H. (1984). Voids in expanding Universe, in *General Relativity and Gravitation*, ed. B. Bertotti, F. de Felice and A. Pascolini. D. Reidel, Dordrecht, pp. 289 – 312. [164, 197]

Sato, H. and Maeda, K. (1983). The expansion law of the void in the expanding Universe, *Progr. Theor. Phys.* **70**, 119. [15]

Schneider, J. and Célérier, M. N. (1999). Models of universe with an inhomogeneous Big Bang singularity II. CMBR dipole anisotropy as a byproduct of a conic Big-Bang singularity, *Astron. Astrophys.* **348**, 25. **[55, 56, 201, 204, 207]**

Schwarz, D. J., Starkman, G. D., Huterer, D. and Copi, C. J. (2004). Is the low-ℓ microwave background cosmic?, *Phys. Rev. Lett.* **93**, 221301. **[200]**

Seljak, U., Sugiyama, N., White, M. and Zaldarriaga, M. (2003). Comparison of cosmological Boltzmann codes: are we ready for high precision cosmology?, *Phys. Rev.* **D68**, 083507. **[189]**

Seljak, U. and Zaldarriaga, M. (1996). A line of sight approach to cosmic microwave background anisotropies, *Astrophys. J.* **469**, 437. **[189]**

Silk, J. (1977). Large-scale inhomogeneity of the Universe: spherically symmetric models, *Astron. Astrophys.* **59**, 53. **[134]**

Simon, J., Verde, L. and Jimenez, R. (2005). Constraints on the redshift dependence of the dark energy potential, *Phys. Rev.* **D71** 123001. **[150]**

Smoot, G. F., Bennett, C. L., Kogut, A., Wright, E. L., Aymon, J., Boggess, N. W., Cheng, E. S., de Amici, G., Gulkis, S., Hauser, M. G., Hinshaw, G., Jackson, P. D., Janssen, M., Kaita, E., Kelsall, T., Keegstra, P., Lineweaver, C., Loewenstein, K., Lubin, P., Mather, J., Meyer, S. S., Moseley, S. H., Murdock, T., Rokke, L., Silverberg, R. F., Tenorio, L., Weiss, R. and Wilkinson, D. T. (1992). Structure in the *COBE* differential microwave radiometer first-year maps, *Astrophys. J.* **396**, L1. **[201, 207]**

Stelmach, J. and Jakacka, I. (2001). Non-homogeneity-driven universe acceleration, *Class. Quant. Grav.* **18**, 2643. **[148]**

Stephani, H., Kramer, D., MacCallum, M., Hoenselaers, C. and Herlt, E. (2003). *Exact solutions of Einstein's field equations*, 2nd edn. Cambridge University Press. **[94]**

Stoeger, R. W., Ellis, G. F. R. and Nel, S. D. (1992). Observational cosmology: III. Exact spherically symmetric dust solutions, *Class. Quant. Grav.* **9**, 509. **[71, 181]**

Stoeger, W. R., Maartens, R. and Ellis, G. F. R. (1995). Proving almost homogeneity of the Universe: an almost Ehlers–Geren–Sachs theorem, *Astrophys. J.*, **443**, 1. **[189, 213]**

Sussman, R. A. (2008). Conditions for a negative 'effective' acceleration in Lemaître–Tolman–Bondi dust models, arXiv:0807.1145 **[147]**

Szekeres, G. (1960). On the singularities of a Riemannian manifold, *Publicationes Mathematicae Debrecen* **7**, 285; reprinted, with historical comments: *Gen. Rel. Grav.* **34**, 1995 (2002). **[13]**

Szekeres, P. (1975a). A class of inhomogeneous cosmological models, *Commun. Math. Phys.* **41**, 55. **[4, 6, 16, 43, 94]**

Szekeres, P. (1975b). Quasispherical gravitational collapse, *Phys. Rev.* **D12**, 2941. **[4, 6, 16, 43, 51, 52, 83, 94]**

Szekeres, P. (1980). Naked singularities, in *Gravitational radiation, collapsed objects and exact solutions*, ed. C. Edwards. Springer (Lecture Notes in Physics, vol. 124), New York, pp. 477 – 487. **[69]**

Tanimoto, M. and Nambu, Y. (2007). Luminosity distance–redshift relation for the LTB solution near the centre, *Class. Quant. Grav.* **24**, 3843. **[39, 40, 179]**

Thompson, K. L. and Vishniac, E. T. (1987). The effect of spherical voids on the microwave background radiation, *Astrophys. J.* **313**, 517. **[189]**

Tolman, R. C. (1934). Effect of inhomogeneity on cosmological models, *Proc. Nat. Acad. Sci. USA* **20**, 169; reprinted, with historical comments: *Gen. Rel. Grav.* **29**, 931 (1997). **[4, 11, 40]**

Tomita, K. (2000). Distances and lensing in cosmological void models, *Astrophys. J* **529**, 38. **[138, 147]**

Tomita, K. (2001). A local void and the accelerating Universe, *Mon. Not. R. Astron. Soc.* **326**, 287. **[147, 148]**

Tomita, K. (2003). Dipole anisotropies of IRAS galaxies and the contribution of a large-scale local void, *Astrophys. J.* **584**, 580. **[147]**

Turner, M. S. (1991). Tilted universe and other remnants of the preinflationary universe, *Phys. Rev.* **D44**, 3737. **[201, 205]**

Unruh, W. G. and Losic, B. (2008). Aspects of nonlinear perturbations in cosmological models, *Class. Quant. Grav.* **25**, 154012. [**3**]

Uzan, J.-P., Clarkson, C. and Ellis, G. F. R. (2008). Time drift of cosmological redshifts as a test of the Copernican principle, *Phys. Rev. Lett.* **100**, 191303. [**173**]

Vale, C. (2005). Local pancake defeats axis of evil, arXiv:astro-ph/0509039. [**200**]

van Elst, H., Uggla, C., Lesame, W.M., Ellis, G.F.R. and Maartens, R. (1997). Integrability of irrotational silent cosmological models, *Class. Quant. Grav.* **14**, 1151. [**3**]

Vanderveld, R. A., Flanagan, E. E. and Wasserman, I. (2006). Mimicking dark energy with Lemaître–Tolman–Bondi models: weak central singularities and critical points, *Phys. Rev.* **D74**, 023506. [**14, 174, 176**]

Wiltshire, D. L. (2007a). Cosmic clocks, cosmic variance and cosmic averages, *New J. Phys.* **9**, 377. [**3, 214**]

Wiltshire, D. L. (2007b). Exact solution to the averaging problem in cosmology, *Phys. Rev. Lett.* **99**, 251101. [**3, 214**]

Yoo, C. M., Kai, T. and Nakao, K-i. (2008).a Solving the inverse problem with inhomogeneous universe, *Progr. Theor. Phys.* **120**, 937–960. [**172**]

Index